Patrick Neill

Patrick Neill,

1776—1851:
Doyen of Scottish Horticulture

by

Forbes W. Robertson

Whittles Publishing

Published by
Whittles Publishing Ltd.,
Dunbeath,
Caithness, KW6 6EG,
Scotland, UK

www.whittlespublishing.com

ISBN 978-184995-032-9

Printed by Melita Press, Malta

Other books by the author

Robertson, F. W. *Early Scottish Gardeners and their Plants, 1650–1750.*
Tuckwell Press: Edinburgh, 2000.

Robertson, F.W. & McKelvie, A. *Scottish Rock Gardening in the Twentieth Century.* Scottish Rock Garden Club, 2000.

For
Alastair, Colin and Fiona

Patrick Neill was the first Secretary of the Society from its foundation in 1809. It remains in good heart with around 100 members from all round Scotland and indeed in England and further afield.

"The Caley" provides leadership to Scottish horticulture: by co-founding and maintaining *Gardening Scotland*; by initiating and supporting the *Scottish Gardeners' Forum*; by making prestigious awards in recognition of distinguished service to Scottish horticulture; and of course by providing activities and facilities to its membership.

New members – whether established gardeners, complete beginners, or somewhere in between – will be very welcome.

Contents

Preface

Patrick Neill's contribution to the development of horticulture and gardening in Scotland has not received the attention it deserves. During the first half of the 19th Century he played an important part in the life of different categories of society in Edinburgh, so it is at first surprising that he has not won greater recognition. There can be little doubt that this omission stems from the loss of his note-books, papers and correspondence which were last heard of about a century ago. So far all efforts to discover whether they still exist have failed. In the absence of such personal information it is impossible to construct a full and coherent account of his dealings with business associates, botanists, gardeners, nurserymen etc – the sort of information on which a biographer relies. Nevertheless, there is sufficient material in the public domain, unnoticed or forgotten, to establish his reputation as a significant figure in the annals of horticulture in Scotland.

As the head of a leading firm of printers, Neill devolved the daily affairs of the business to an able and loyal manager (and eventual partner) leaving him free to pursue his interests in horticulture, gardening, natural history and antiquities. His botanist friends included some of the most eminent scientists of the day. The Royal Caledonian Horticultural Society owed a great deal to his devoted service as secretary for some 40 years. His publications, which are models of careful reporting, are of considerable historical importance. Often providing a revealing picture of contemporary events and attitudes, they are the principal themes of the present account.

Forbes W. Robertson, Edinburgh

Acknowledgements

I wish to thank the Royal Caledonian Horticultural Society and the Binks Trust for their generous financial support. I thank John Ballantyne for his useful Shetland references.

1 🌿 *Perspectives and antecedents*

Patrick Neill (1776–1851) was a man of many parts. Head of the largest printing firm in Edinburgh, he contrived to delegate the daily running of the business to a devoted manager and eventual partner, leaving himself free to pursue his major interests in horticulture, botany and natural history. He was also an antiquarian, a town councillor, and an active member of the Church of Scotland, serving both as elder and periodically as a lay member of the General Assembly. Leading members of the academic fraternity and the professions were numbered among his friends. He was a founder member of both the Caledonian (later Royal) Horticultural Society and the Wernerian Natural History Society, which in its heyday rivalled in prestige the Royal Society of Edinburgh (of which he was also a Fellow). For the greater part of his adult life he served as Secretary to both the Caledonian and Wernerian Societies, contributing substantially to the intellectual life and wellbeing of his native city.

Although his interests ranged widely, it is his devotion to horticulture, botany, gardening and natural history which are our present concern. At his death he left behind notebooks, unpublished manuscripts, correspondence and papers which he bequeathed to his long standing friend, the Rev. John Fleming, Professor of Natural Science in the Free College, Edinburgh. Fleming died not long after Neill, whose Executors sought to persuade the Professor of Botany at Edinburgh University, John Hutton Balfour, to write a biography of Neill.[1] However the appeal came to naught. The notebooks and papers were never deposited in a public archive. Just over a century ago they were reported to have vanished and all efforts to discover whether they might still exist have since proved fruitless.[2] Consequently a great deal of information about Neill's life, his youth and the dealings he had with so many people in different trades and professions – all grist to a biographer – are lost to us. But fortunately, his publications, valuable correspondence that has survived, as well as contemporary reports and records together reveal the distinguished part he played in the development of Scottish horticulture and gardening.

Patrick Neill was descended from a family of Haddington printers. The founder of the printing firm of Neill & Company – to which he succeeded – was his uncle, also called Patrick. In 1725, at the age of 14, Neill's uncle left home to serve his apprenticeship with the Edinburgh printer James Cockburn. He evidently thrived in his profession, for in 1749 he entered into partnership with two booksellers: Gavin Hamilton and John Balfour. The former was the son of Dr William Hamilton, the Principal of Edinburgh University. The publisher Archibald Constable considered the firm to be the most respectable retail business in Edinburgh. The three partners quickly established a reputation for producing high quality work. Many notable volumes were printed, including David Hume's *Essay on History*, editions of Virgil and Horace, Terence's *Comedies* and the works of English poets, including Pope. The firm was appointed printers to the University.[3] [4]

The partnership lasted for 16 years. When Hamilton and Balfour retired, Patrick introduced as partner his brother Adam, the father of our Patrick Neill. The printing firm was at first located on the premises of Edinburgh University – or College as it was then known – but in 1769 it was transferred to Old Fishmarket Close, a precinct off the High Street formerly favoured by wealthy citizens. This was a strategic location, close to the firm's major customers, the College and the Law Courts. About 1778 Patrick retired to the country and Adam Neill took on Robert Fleming, a bookseller, as partner. But this arrangement did not last long and shortly thereafter Patrick's son James entered the business as partner. In 1802 Adam's son Patrick – the subject our story – was brought in as a third partner at the age of twenty-six. In 1808 James Neill bought a farm at Foxhall in what proved to be a financial disaster that mired him in debt. In 1812 Adam Neill died, leaving the two cousins James and Patrick in charge of the firm. But they soon quarrelled and the firm was split into two.[5] The pros and cons of the dispute between the cousins is obscure. In a letter to his lawyer, James complained that although he had worked harmoniously for many years with his uncle Adam, as soon as he died, Patrick adopted a 'wrangling disposition', bringing in an accountant to go through the books and seeking the advice of his own 'man of business'.[6] It is possible that this was prompted by concern that James was taking too much out of the profits to pay for his debts. James also complained that Patrick, although young and able, never spent more than an hour or two on his daily visits to the premises; an observation which rings true as early evidence of the competing attractions of his other interests, which increasingly claimed most of his attention. The upshot of the dispute was that James abandoned his Edinburgh concerns and later emigrated, leaving Patrick in sole charge of Neill & Company.

Neill was evidently both a competent printer and businessman, for the firm prospered under his control. Although it provided a comfortable income, he was loath to devote the time required for the day-to-day running of the business, since he was so fully committed to both the Caledonian Horticultural and the Wernerian Natural History Societies, as well as his botanising and his garden. This problem was solved when Neill employed an assistant, William Fraser. Fraser became manager, handling all the day-to-day business, and eventually Neill's partner, although the original name of the firm was retained. He was responsible for modernising the plant with steam driven presses and was in all respects, reliable and competent. There is little doubt that he felt a debt of gratitude to Neill for devolving so much authority to him, for he called one of his sons Patrick Neill Fraser. The firm continued to print scientific works, Government and legal publications, editions of the *Encyclopaedia Britannica* and other prestigious works. Fraser died suddenly and unexpectedly in 1846 and Neill found himself shouldering the responsibility of running the firm at a time when he was on the point of formal and effective retirement. However, he hastened to bring in Fraser's sons, Alexander and Patrick Neill Fraser, who took over the business when Neill died in 1851 and ran it with great success.[7]

Little can be said of Neill's early life. Apparently he was brought up as an Anti-Burgher, an extreme Protestant sect. However, he later adhered to the established Church of Scotland and stayed with it at the Disruption of 1843.[8] He became an elder of St Mary's Kirk and was, on several occasions, a lay member of the General Assembly, representing the Northern Isles of Orkney, where he had relatives. He was no bigot and remained a close friend of some of the dissenters. Since his name does not appear among the pupils attending either the High School or Heriots Hospital, it is not known where he received his early education. It was his original intention to train as a surgeon at the University of Edinburgh. The matriculation records of the University show that Patricius Neill was in Andrew Dalzell's Latin class in 1789 and 1790. He must have been an attentive pupil, since in later life he was never at loss for an apt quotation from the classics. The extent of his medical studies is unknown but if he attended medical classes he would have been taught botany, which was an essential constituent of a medical training, and this may have laid the foundations for his life-long interest in plants. He soon became a competent field botanist, familiar with the wild flora of Scotland. When he inherited Canonmills Cottage and garden he established a reputation as collector of newly introduced species of plants from many different parts of the world.

He never graduated, for it seems his university career was cut short by his joining the firm, no doubt meeting the wishes of his father who needed assistance in coping with the increasing demands of the expanding business, and perhaps also because cousin James was not pulling his weight. Given Neill's devotion to science and learning, the termination of his studies must have been an acute disappointment. Indeed, he was something of an academic manqué. Many of his closest friends were university professors and lecturers. Since his firm printed so many academic works the premises became a congenial meeting place for the authors and their friends.[9] No doubt Neill revelled in such intellectual company. When a town councillor, he played his part in the partisan campaigns that attended the filling of University Chairs. His subsequent Fellowships of the Royal Society of Edinburgh and the Linnaean Society, and his award of a Doctorate of Laws in 1834 from the University of Aberdeen no doubt compensated for not having a university degree.

We might wonder about his appearance and his character. We learn from several obituaries that he was of moderate stature and of a lean build. His portrait by Patrick Syme shows him as clean shaven, with a prominent nose and chin and an expression which some found rather sardonic. He wore a brown wig, latterly at least. He always dressed in dark clothes with a long black coat and wore a wide white cravat. In his later years he walked with a slight limp. He became a familiar Edinburgh figure, walking up Pitt Street from his home in Canonmills to his premises off the High Street, with a gold-topped cane and a large umbrella.[10] [11]

His good nature and courteous, helpful manner were widely acknowledged. That he served so long as Secretary to both the Caledonian Horticultural and the Wernerian Societies is sufficient proof of his tact and discretion. The Caledonian Horticultural Society expressed its appreciation at intervals by voting him a piece of plate, commissioning his portrait in oils and arranging for a bust by the finest sculptor in Scotland. Near the end of his career he was presented with an elaborate piece of silver plate paid for by subscriptions from 600 Scottish gardeners. His contemporaries held him in the highest regard, none more so than the gardeners of the Scottish lowland estates. The benign exterior however, concealed a streak of uncompromising pugnacity that was revealed whenever he was challenged on some question of principle or his interests were threatened.

Neill was a bachelor whose household was run by Ann Neill, the daughter of his cousin James. The only hint of a possible romantic attachment appeared in his will, in which he bequeathed ten pounds to Marianne Kerr, High Street, Burnt Island, as a 'remembrancer'.

No doubt, once he had taken possession of the family home at Canonmills Cottage, like many bachelors, the routine of his life changed little from day to day. He was very hospitable, frequently inviting friends, fellow botanists and academics, ministers of the Kirk or the Lord Provost for dinner. If any distinguished naturalist visited Edinburgh it would not be long before he found himself dining with Neill, probably in the company of several professional men of similar interests, invited to meet him. Politically, Neill was constitutionally conservative and, as a Tory in a predominantly Whig Town Council, he represented the so-called aristocratic ward of the New Town for many years.

We get a glimpse of life at Canonmills Cottage from the reminiscences of George Bentham, the botanist, who visited Edinburgh in 1823 and again in 1827.[12] On both occasions Bentham noted the wealth of Neill's garden in rare or unusual plants, tastefully distributed in beds and borders. He marvelled at the unexpected collection of birds which included a heron, a cormorant, a wild goose, a solan goose or gannet, a great black backed gull, as well as other species of gull, ducks, cats and dogs, all living harmoniously together and ever ready to be fed. Apart from the birds, there was even an ichneumon and a siren from the swamps of Carolina. At his first visit, his fellow guests at dinner included several of Neill's close friends: Robert Bald, a mining engineer from Alloa; Robert Stevenson, the lighthouse engineer – an almost exact contemporary of Neill and grandfather of Robert Louis Stevenson – and George Arnott, later Professor of Botany at the University of Glasgow. Four years later saw Bentham again Neill's guest at dinner, accompanied this time by the same George Arnott and also his brother. Bentham reflected how, sitting in the same room with much the same company, he experienced a sense of timelessness as if this dinner was but a continuation of the last one he had enjoyed at Canonmills Cottage. He also remarked rather enigmatically, that he found Neill a rather curious kind of man, but sadly did not elaborate further. Bentham commented on the unusual countenance of the household 'factotum'; Peggy Oliver who had a winning way with the little menagerie of animals that followed her about hoping for food. There were also one or two other domestics and, of course, the gardener or rather succession of gardeners who, after a spell with Neill, generally moved on to responsible jobs elsewhere. Thus, we can envisage the little ménage at Canonmills Cottage pursuing a very tranquil and pleasant existence over the years.

Neill was well off since his printing firm was flourishing. Having devolved most of the worries of business to his manager, he was free to pursue his scholarly interests and supervise

Canonmills Loch, Old and New Edinburgh, James Grant, 1880, Cassell

his plant collection. In his last years he was handicapped by failing eyesight and suffered a severe stroke shortly before he died. But he kept up his correspondence and cherished his plants to the very end.

Neill lived through momentous times which shaped the course of history. He was born in the same year that Adam Smith published 'An enquiry into the nature and causes of the wealth of nations'. At the age of 13 he would have heard of the fall of the Bastille in Paris and four years later no doubt shared the horror of the British establishment at news of the execution of Louis XVI and his queen. Then there was the interminable war with France and the threat of a Napoleonic invasion. It was not until he was almost 40 that the Battle of Waterloo ushered in an era of comparative peace. His reaction to these events would probably have appeared in the missing papers. We do know that in 1817 he and two gardener colleagues visited the battlefield of Waterloo and remembered their fellow countrymen who had fallen in the battle. How far the sequence of events in France contributed to his conservative politics is unknown but quite likely. He had reservations about the advantages of the Act of Union of 1707 and had a nostalgic regard for the old Scots ways which were fast disappearing.

1   James McNab's Scrapbook. Library, Royal Botanic Garden Edinburgh.
2   Druce, G. C., *The Life and Work of George Don, 1903–1908*, Notes, Royal Botanic Garden Edinburgh, (3), pp.55–91.
3   Neil & Co., *The Printing House of Neill*, 1917, Paul's Wark, 212 Causewayside, Edinburgh.
4   Moray McLaren (ed), 1949, *The House of Neill, 1749–1949*, Neill & Co., Edinburgh.
5   *Ibid.*
6   Gen 29/182. LA/75. National Library of Scotland.
7   *The Printing House of Neill.*
8   J.B. Barclay, 1852, *Patrick Neill, LL.D, FRSE, Oct. 25, 1176–Sept. 3, 1851. The Cottage Gardener*, (VII), p.121.
9   Moray McLaren, *The House of Neill.*
10   J.B. Barclay, *Patrick Neill.*
11   P.P. Brotherstone, 1923, *Patrick Neill, Gardeners Chronicle*, vol. *LXIII*, 3rd series, pp.320–321.
12   Marion Filipuik, (ed), 1992, *George Bentham, 1810–1834*, University of Toronto Press, p.158, p.165.

An Edinburgh 'character'; Patrick Neill in later years, B.W. Crombie, Modern Athenians, 1882, Adam & Charles Black, Edinburgh.

Portrait of Sir William Jackson Hooker.
By permission of the Royal Botanic
Gardens, Kew.

Portrait of Robert Brown. By permission
of the Royal Botanic Gardens, Kew.

Cartoon of George Combe, Edinburgh
lawyer and leading phrenologist, B.W
Crombie, *Modern Athenians*, 1882,
Adam & Charles Black, Edinburgh

2 ❦ The Antiquary

The earliest of Neill's writing that has survived refers to one of his subsidiary interests so we had better start with that. The buildings and artefacts of a bygone age excited his interest from an early age and never left him. In later life he strove – with success – to preserve the landmarks of old Edinburgh, threatened by civic improvers. In due course he was elected to the Council of the Society of Antiquaries. In 1799 when he was 23 years old, he joined with two or three kindred spirits of similar age to form a little Society of Antiquaries. The cardboard record book bore the ponderous title: 'Journal of Different Excursions made by individuals of a Society of Antiquaries chiefly for the purpose of investigating Monuments of Antiquity in Scotland. Edinburgh 1799'.[1] The first entry recorded how 'this evening the Society of Antiquaries met in the Covenant Room, Covenant House, now Mrs Martin's Tavern, where they resolved themselves into a society for the purpose of writing Journals of Visits to Antique buildings and other Monuments of Antiquity in Scotland and occasionally making observations of the natural curiosities.' In accordance with the contemporary zeal for orderly procedure, they gravely formulated a set of rules, as follows:

1. That every paper in the Journal shall be read before the members of the society and approved by them before insertion in the Book.

2. That the Journal Book shall be kept by the members successively, and keeping it for a quarter of a year, but that it shall always be open to inspection of the other members.

3. That every paper to be inserted shall have an appropriate motto.

4. That the society will meet as often as appears convenient, the meeting to be always in some antique or singular place.

5. That each meeting of the society shall be mentioned in an appropriate minute.

At the inaugural meeting there was produced and read to the society the petition to the Magistrates of Edinburgh which, it was agreed, should be transmitted to the Lord Provost the next day. The embryonic society took itself very seriously. This formal statement was signed by Pat. Neill, James Walker and John Stark. The reports of two of the founding members who had already written up their visits were read, approved for insertion in the Society's Book, and the society was launched. There was a post-script to the effect that the first volume of the Transactions of the Royal Society of Antiquaries was purchased for 6s.6d in Ross's Sale Room, as the first contribution to their projected library.

The first formal meeting of the society took place on March 12 1800, when all three members met – as per rule – in the ancient precincts of Craigmillar Castle. Here they added a further injunction to the rules, namely that the Society's Book or any other book or things that may be purchased or presented to the society should never be removed from Edinburgh; that in case of any members leaving Edinburgh, the resident members should always have the control of record books etc. which would become the property of the society members. And then, with disarming modesty, they decided that the members should use every endeavour to publish or otherwise hand down to posterity 'this small production of three juvenile antiquaries'.

The Book of the Society contains about a dozen reports of visits to different sites by one or other of the members, either alone or in company with the others, occasionally including an additional Mr McGregor. These accounts, written in a fair hand and introduced by a suitable quotation – often a verse or two from a classical author – include quite detailed descriptions of buildings and tombs. Since most of these have often been described elsewhere or may be familiar to the reader, there is no point in repeating them here and we are therefore content to note the destinations of Neill's excursions and any incidental observations of particular interest, especially if they refer to some long forgotten aspect of the Scottish scene.

Neill's first record refers to a trip to Fife. Leaving home at six in the morning on September 14 1799 he made for Leith, where he boarded *The Diligence* passage boat to Kirkcaldy. On the way he saw for the first time the oyster dredgers at work and heard them bawling the nonsensical songs that served to regulate the time allotted to pulling the nets. He was soon becalmed and had to transfer to a small fishing skiff to reach the shore at Kinghorn. After breakfast at a local hostelry he proceeded to Kirkcaldy, noting an exposure of fossil-bearing rocks and a deep cave on the way. After dining 'at Wilson's' he visited Ravenscraig Castle and later reached the New Inn, which he rated a good 'house of entertainment'. We could infer that Neill was something of a *bon viveur* for he rarely missed mention of the inns he visited. Since he came from a well-heeled family, he was under no restraint to go hungry. His next objective was via Cupar to St. Andrews where he remarked on the old buildings as he walked to the ruins of the cathedral. He contemplated the habitation of the infamous Beaton and saw the window from which, according to local tradition, the Cardinal 'glutted his eyes with the burning of the pious Wishart'. As a good Presbyterian, Neill relished how retribution came to the Cardinal on May 29 1546. From St. Andrews he walked to Leuchars and thence to Woodhouse where he crossed the 'Frith of Tay' by ferry to Dundee en route for Perth and Scone. At Elcho Castle he turned for home via Abernethy where he examined the Pictish tower and Falkland Palace. Here he found the ruins occupied by the parish minister; the sole occupant. From Falkland he made for the

New Inn and then via Balbernie Castle to Kirkcaldy and Pettycur, where he immediately embarked for Leith. Neill was evidently a stout walker, having travelled from Elcho Castle – not far from Perth – in a day.

The next entry referred to an earlier expedition on November 29 1798, before the formal establishment of the society. Neill, Walker, Stark and McGregor set off to visit Roslin, going rather indirectly along the foot of the Pentlands. Just south of Edinburgh after passing a Roman camp, they encountered a large, black, seven-foot whinstone in a stubble field. It went by various names such as Kel-stone; claimed to be an ancient British compound name signifying Battle Stone. It was also referred to as the Camus Stane, although the local people knew it as the Buck Stane; traditionally the place where the king and nobles assembled for the start of the chase. They then went over the Pentlands via Swanston – with 'no small danger to their limbs' since the going was slippery – and soon reached Rosslyn chapel While dinner was being prepared 'at Wilson's', they visited Rosslyn Castle. On the way home by Loanhead, they were put to some inconvenience by the dirtiness of the roads and the 'darkening of the night', reaching home about eight o'clock.

On September 14 1799 Neill set off on a solitary 'pedestrian' trip to Linlithgow, Falkirk, Stirling, Glasgow and the Falls of Clyde, leaving Edinburgh by Corstorphine, 'famed for its cream'. Finding the kirk door open he explored the interior. Kirkliston was reached about noon so he took refreshment at Mackay's Inn, where he noted the date of 1692 above the door. He reached Linlithgow about three in the afternoon, described the well-known features and pressed on to reach Falkirk by eight – not bad he thought, for a day's walking, considering that the roads were very wet and indeed, 'a perfect puddle' caused by the great droves of cattle coming from the Falkirk Tryst held earlier in the week.

Next day he crossed the River Carron by an excellent stone bridge, passed by Bannockburn where he expressed his patriotic sentiments, and thence on to the ruins of St Ninian's church, which had blown up while serving as an ammunition store during the Forty Five. Then he made for Stirling – described in some detail – and Glasgow where he was greatly impressed by the cathedral. From there he made for Bothwell and soon crossed the Cawdor near its junction with the Clyde. Here he noted in nearby woods, the moss *Hypnum repandum* and the stink horn fungus. After dining at The Gilded Stag's Head Inn he set off for Lanark, noting that the country abounded with orchards. The banks of the Clyde were covered with fruit trees. At Dalserf he had a fine view of the modern castle of Maudslie. Lanark – invariably pronounced 'Lenrick' by the inhabitants – held little of interest for him compared with New Lanark and ultimately, the Falls of Clyde. Here, after paying the entrance fee, he declined the services of the guide since he wished to contemplate the impressive spectacle by himself. He left this 'uncommon scene with regret' and turned for home via the Peebles road. Near Linton, after wading the West Water in spate and regretting the lack of a bridge, he made a beeline for an armchair before a warm peat fire at the good Peter Brook's inn. Neill called him 'good' because he was not ashamed to open the 'Ha Bible' and 'patriarch-like' exclaim before his family: 'Let us worship God'. He left this pious publican's abode the following morning, to breakfast at Nine Mile Burn, but found it a poor place quite in keeping with the surrounding country. Formerly it had been reckoned to be nine miles from Edinburgh, but was now measured as twelve. Nothing remarkable occurred on his way home.

On April 9 1800 he accompanied 'that acute and indefatigable student of nature, Mr John Mackay' – who was then the superintendent of the Royal Botanic Garden – on a botanical expedition to the islands of Inchkeith and Inchcolm. They set off, well equipped with bread, cheese, crabs, porter and whisky, to venture on the Forth in one of the small Newhaven fishing boats. At first the sea was like a mirror but near Inchkeith an east wind got up. They landed on the east side of the island on ledges of rock exposed by the ebb tide, where they saw *Fucus pinnatifidus* (or pepper dulse) and large sea-urchins. The uninhabited island had a few good springs and a decent pasture. There was a stone monument where some thirty Russian seamen were buried, having perished of a contagious fever while their ships were lying in the Leith Roads. There was a rough stone at the head of each grave. While on their way to Inchcolm, one of the company produced a gun and shot a black guillemot and a sea-swallow (i.e tern). On Incholm they found several interesting plants including deadly nightshade, common horehound, sea lovage, henbane, shepherd's rod, rape and an unidentified *Brassica* – not in flower – taken back to be planted in the Botanic Garden for future identification. The only tree or shrub on the island was the bour tree or elder, which appeared indigenous. They found an immense amount of pellitory-of-the-wall on the abbey walls. Their trip back to Newhaven was over the deepest part of the Forth, known as Mortimer's Deep. The sea was now boisterous and Neill asked the fishermen whether this was not stormy. 'Oh no' they replied, 'it is fine weather'; sentiments not shared by the two passengers who were drenched with spray and seasick. To their relief, in about an hour and a half, they reached Newhaven where they made for the nearest hostelry and a dish of tea, which they took while viewing the billows through the window, glad to be on dry land.

On May 12 1800 Neill described his visit to the Palace and Abbey of Holyrood. He was shown around the former by a succession of guides, four in all. He was directed to observe 'what was religiously inculcated upon all visitors, the identical stains imprinted on the floor of the purple tide of the murdered Rizio'. He received this information with appropriate scepticism.

On September 30 1800 Neill made a trip to southern Scotland via Glasgow, Port Glasgow and Greenock, where he found the roads so 'disagreeably miry' that he hired a 'taxed cart' belonging to one Francy Tarbert, who told him that the same machine had conveyed the painter Nasmyth when he was drawing his views of Ayrshire. Finding Largs and Kilbride equally poor he went on to Kilwinning Abbey. He claimed that it was here, during the building of the Abbey, that free-masonry was introduced to Scotland by foreign artists. Considering the piety and reverential superstitions of those times, he thought it highly probable that free masonry could not have countenanced the despicably puerile and so daringly blasphemous ceremonies that had been revealed at Ayr. Archery was still practised here. The archers annually shot at a target fixed to the head of the steeple. Between Kilmarnock and Irvine he was passed by carts carrying a sort of coal, like charcoal, which was exported for manufactures in Ireland. He completed the journey from Kilmarnock to Glasgow cooped up in a chaise. He was unlucky with the weather for he met with equinoctial gales and torrents of rain, occasionally interrupted by periods of sunshine which, by contrast, he 'enjoyed with superior relish'.

The accounts of the various expeditions by members of the society were followed by a few corrections of earlier statements. Thus he had mistakenly assumed that a tomb of Lord Belhaven marked the remains of a contemporary of Fletcher, whereas it proved to be the tomb of a 17th Century ancestor. He added, a little sanctimoniously: 'May members of the society never be ashamed, cordially to acknowledge and rectify their mistakes.' And there the records of the society end, for the rest of the dozen or so pages of the Book are blank. Like so many youthful enterprises, the initial enthusiasm may have waned or Walker and Stark may have left Edinburgh – Neill could hardly comprise a society on his own. But at least their aim of leaving their records to posterity has not proved wholly in vain.

Although the little society faded away, later in life he sought to preserve some of the landmarks of old Edinburgh. Thus in 1829 he campaigned by pamphlet for retention of the Flodden Wall and Tower.[2] He observed that the construction of the wall was started in 1513 immediately after the disaster of Flodden, where the Lord Provost and several members of the town council had taken the field. On news of the disaster, when a contagious epidemic was raging in the city, the town council at once issued a proclamation commanding all male citizens to assemble, bearing arms, at the tolling of the common bell while the women were to repair to the church and pray for success. The council next looked to the fortifications of the city which had extended beyond the old wall, so that the Grassmarket and Cowgate – which had become fashionable places to live – were exposed to attack. A new wall was started from the corner of Castle Rock to the West Port, ascending in line of the steep lane known as the Vennel and turning eastward north of Heriot's Hospital. Here there was a small fortress or tower, with loop-holes designed to cover with gunfire approach to the West Port. With the passage of time, most of the wall had been destroyed and only the tower and the part by Heriot's Hospital remained. Neill argued that the Vennel Tower should be deemed sacred since it was a prized monument of the invincible spirit of the Edinburghers in their struggle against the English. A number of residents had petitioned in favour of demolition of the tower, claiming it was a lurking place for thieves and hence a nuisance. But Neill claimed they must have known little of its history and historical significance. A few modifications and planting with evergreens would eliminate the alleged nuisance. Neill added a lengthy footnote worth comment since it reveals his patriotic sentiments and reservations about the Act of Union and the need to preserve all Scottish rights and institutions recognised therein, and guard against loss of national virtues ands even national religion. Never was it more important to foster the Scottish spirit than just then. That which force of arms had failed to achieve was being won by gentler means. He got quite carried away. If erosion of traditional values and customs were continued, the Lord Provost might yet be called on to evince the spirit of a George of Tours and the Convener to unfurl the Blue Blanket. In the face of this call, almost to arms, the petition for the demolition of the Tower must have appeared a tawdry gaffe. Neill carried the day and both the tower and the surviving part of the Flodden Wall have become valued relics of old Edinburgh.

In 1837 and 1838 he opposed proposed alterations to Parliament Square and the Market Cross. If carried out they would exclude access to certain closes and could lead to the erection of buildings on space better left open. Barring entry to one of the closes in question would prevent ready and traditional access to his printing firm in Old Fishmarket Close. In a pamphlet published jointly with his partner William Fraser, it was argued that the whole area from Old Fishmarket Close to St Giles should be left as an open space.[3] They strongly opposed the suggested loss of closes. It was necessary to explain to an Englishman that a 'close' was not a cul-de-sac but a subordinate thoroughfare. The pamphlet appealed to the Parliamentary Trustees – instigators of the proposals – to acquaint themselves with the views of the town council, the Merchants' Company and the Commissioners for the Improvement of Edinburgh. Neill, who was then a member of the town council, continued his campaign at successive – often acrimonious for opinion was divided – meetings of the council. However in August 1837, a memorial was sent to the Lords of the Treasury opposing the disposal of the vacant space about the Market Cross, sparking contentious sessions among the councillors who were divided on political grounds.[4] Eventually Neill's intervention was rewarded, even if he did not obtain all that he campaigned for. His efforts to protect old landmarks from the improvers won the approval of Lord Cockburn, who accorded him the title of 'useful citizen'.

1   Acc. 6285 *Journal of Different Excursions Made by Individuals of a Society of Antiquaries Chiefly for the Purpose of Investigating Monuments of Antiquity in Scotland, Edinburgh 1799.* The National Library of Scotland.
2   YDA 1876, *Notes relating to the Fortified Walls of Edinburgh, with a Copy of the Proclamation Issued by the Town Council on Receiving the News of the Battle of Flodden,* West Register House, Edinburgh.
3   Patrick Neill and William Fraser, *Observations on the Proposal of Building at the Cross, and of Shutting up the Royal Bank Close,* 1837, The National Library of Scotland.
4   Patrick Neill, *Observations on the Proposal of Building at the Cross, and of Shutting up the Royal Bank Close. With Remarks on a Report of the Lord Provost's Committee, 19th July 1837; and an Account of Subsequent Proceedings in April 1838, Relative to the Subject,* 1838, The National Library of Scotland.

3 🌿 A Tour of the Orkney and Shetland Islands, 1804

In the summer of 1804 at the age of 28, Neill visited Orkney and Shetland. He had relatives in Stronsay, Orkney, although he did not refer to them in the account of his trip which was published as a number of instalments *in The Scots Magazine* of 1805 and 1806. Later they were combined and published as a book in 1806, with the unwieldy title: 'A Tour through some of the Islands of Orkney and Shetland, with a view chiefly to Objects of Natural History but including also occasional remarks on the State of the Inhabitants, their Husbandry and Fisheries' – brevity of expression was not Neill's style.[1] This book first brought his name to public notice, or at least the well-educated readers of *The Scots Magazine*. The instalments and the book created interest on two counts. Firstly, Neill as naturalist and antiquarian, wrote about the animals, plants and natural features of the Islands and in such respects, the book retains its historical value. Secondly, appalled by what he took to be the unhappy state of the Shetland islanders derived from their excessive dependence on their landlords, he attacked this apparent social evil in uncompromising fashion. A spokesman for the landlords, writing under the pseudonym of 'Thule', indignantly replied and attacked Neill by pamphlet and newspaper. He became embroiled with the publishers and editor of the magazine, who sought to curtail the length of his replies and were not enamoured of Thule's invective style[2] [3] [4] [5]. Neill was under no obligation to comment on the political and social scene. He could have confined his account to natural history, agriculture and people's habits, like other travel writers, but he was really quite angry at what he encountered and replied to the attacks upon him with biting sarcasm, not budging an inch. When the instalments were combined as a book – on the principle of *Audi alteram partem* – he thought it proper to include his critic's observations and his own replies so that the reader could judge for himself. It is likely that the sympathies of the readers of *The Scots Magazine* were with Neill. A reviewer remarked on the service he had performed in publicising the unfortunate position of the Shetland islanders and regretted the obloquy he had incurred.

On July 17 1804 Neill set off from Leith in a sloop bound for Thurso and by the 20th was anchored off the town in the Scrabster Road. Conspicuous in the treeless landscape, Sir John Sinclair's house was the first noticeable feature in the bay. When Neill landed, he discovered it was market day in Thurso. He saw Highlanders from the hills of Caithness and from Sutherland dancing the 'fling' in the streets to the music of bagpipes. The following day he crossed the Pentland Firth in an open boat in a heavy swell, passing the islands of Faral and Cava where boats congregated to fish for the speckled dogfish, known locally as 'dogs' or 'hoes'. Oil was extracted from the liver, leaving the rest of the body to be dried for winter food. At mid-day they sailed up Scapa Bay with St Magnus Cathedral visible in the distance. He gave a thumbnail description of Kirkwall, which consisted essentially of an often narrow, mile-long street that was in many places dirty, although the cathedral was open and clean. In the cathedral his botanist's eye noted that the roof and pillars were decorated with outlines of foliage that resembled the designs on the walls and pillars of Rosslyn Chapel outside Edinburgh. A little to the north of the town he found the remains of the fort constructed in Cromwell's time, together with heavily rusted six or eight pounder cannon. He was surprised to find that there was no market for butcher meat, poultry or even fish. Starlings were everywhere in Kirkwall, as common as sparrows in Edinburgh. Among the public buildings he noted the New Church; a large meeting house belonging to the Anti-Burgher Seceders. The preacher was popular, so he usually had a large audience in the spacious building. The only other place of worship belonged to the Established Church.

Neill peered into the gardens and found artichokes growing luxuriantly, as well as cabbages and cauliflowers. As proof of the mildness of the climate he saw 'in blow' sweet marjoram, mignonette, loosestrife, asters, creeping polemonium and tobacco plants more than two feet high, which had been planted as seed in the open border that same season. Elecampane (*Inula helenium*) was commonly grown to provide a decoction which was often added to Orkney ale. In several gardens there were 20–30 foot trees; generally ash and planes (sycamores).

Since the introduction of the kelp industry there had been great changes in the state of society. Country gentlemen gained from their bleak estates riches far beyond what they had known before. Many had abandoned their lonely dwellings – especially during the winter – for the lively, social life and attractions of Kirkwall. In spite of it being a centre of trade, the town lacked a quay or harbour to provide a landing for visitors, who had to either wade ashore or be carried on the backs of the boatmen.

A few miles north west of Kirkwall, at Quanterness, there was what was called a 'pecht's house' – evidently an earth-house. About a mile from the town there was the charming Bay of Scapa, into which flowed a small stream frequented by many large sea trout that appeared to be of no interest to the inhabitants. Neill, ever the moralist, thought it 'foolish supineness' to be so indifferent to such a potential source of food. About the middle of August a shoal of herring appeared in the bay, but only a single boat without a suitable net from Kirkwall put to sea in pursuit, although a couple of boats from Thurso turned up at the site. Neill reflected on the irony of how often he met islanders who were hard pressed for food, yet lived by the sea which could provide them with plenty. At Scapa Bay he noted the presence of the short eared owl as well as a tame racoon which roamed the shore. A Doctor Sutherland – a pupil of Boerhave – had visited a small glen called the Guills of Scalpa, to gather simples for use in his medical practice. Neill could not resist having a look. There he found the common valerian, wild angelica, eyebright, kidney vetch, purging flax, lady's smock and an occasional *Orchis latifolia*. He found bog bean in a marsh and the Scots lovage growing on the rocky shore.

The only serviceable road in Orkney ran from Kirkwall to the point of Holme (pronounced 'Ham' locally), regularly traversed by the post-boy on foot. There was a sort of road from Kirkwall to Stromness but in winter the frequent small streams needed bridging, while the outlet from the Lake of Stenness needed a proper bridge. In general, Neill felt the Orcadians merited government assistance to improve their lot and clothe their children adequately. Such aid would very likely bring returns in increased recruitment to the navy.

He did not approve of much of the islanders' agriculture, which was generally neglected in favour of the kelp harvest. There was a dearth of winter feed for the cattle and in many places the ground was scratched rather than ploughed. Alternate crops of oats and six or eight rowed barley (*Hordeum vulgare*) – known locally as 'bigg' and 'bere' in mainland Scotland – were grown in successive years, so the exhausted soil only produced a scanty crop by dint of applying rotted sea-weed. The oat species differed from that grown in mainland Scotland. In Orkney they grew the black oat (*Avena strigosa*), easily recognised by its many awns, in place of the white oat (*A. sativa*) of the Lothians. Mixed with the black oat, he often saw the tall 'hygrometric' oat (*A. fatua*). A variety of *A. strigosa*, called Red oat, was also grown. The fields were generally very weedy, with plenty of spurrey, bugloss and corn marigold. The latter was known in central and southern Scotland as 'gules' or 'guilds', long recognised as a weed to be eliminated by a combination of manuring, summer fallow and ploughing. The tall oat grass grew wild everywhere. It could have been turned to use as fodder – as in Sweden – but was never cultivated.

On July 25th Neill rode to Deerness where he saw several species of willow (*Salix*) not more than two feet high, rather dwarfed by common mugwort. Here he learnt that strong ropes used for different purposes were made from the shoots of the crowberry heath (*Empetrum nigrum*), while ropes for suspending baskets or 'caseys' over horses' backs were made from the fibrous roots of the sea reed (*Phragmites australis*). Tethers and bridles were constructed from long meadow grasses such as *Holcus lanatus*, known locally as 'pounce' or 'puns'. Neill was interested in local names which he often noted. At the Bay of Firth in Damsay Sound, three or four miles from Kirkwall, there were good oysters that were not dredged, but gathered from ledges at low tide with the aid of long tongs. They fetched from one shilling to one shilling and sixpence per hundred. At Shapinsay, three miles north of Kirkwall, a Colonel Balfour had established a real garden and installed the first greenhouse and 'stove' or heated greenhouse in Orkney.

Stromness – the principal port for whalers and the Hudson's Bay fleet – provided anchorage for many foreign vessels. He thought the houses in the town were distributed with whimsical irregularity. As an antiquary, he was greatly interested in the Standing Stones of Stenness, which he described in some detail and on which he collected three species of lichen. There was one separate, solitary stone of great size with a large hole in it. There was a local tradition that if a lover and his lass grasped hands through the hole that would seal a sacred vow called 'The Promise of Odin'. The superstitious believed that if a young person put their head through it, then palsy would not afflict them in old age.

He next made for Stronsay, where after a turbulent crossing he found vast flocks of golden plover, which darkened the sky, whilst his ears were assailed by the cries of the chaldrick (oyster catcher) and picketarnie (common and/or arctic tern). Two species of plantain were abundant and provided food for the sheep which were believed to produce the best tasting mutton on this diet. Here he first saw the small enclosures or 'bughts' used for rearing cabbage. These 'plantiecruies' as they were also called, consisted of small patches of earth surrounded by stone walls and were always located near the shore. He noted that the long leaves of the yellow flag were used as a coarse sort of hay.

Sailing along the shore of Stronsay he saw many scarfs or cormorants on the rocks. The boatmen, using hooks, dragged down some of the young birds, which were very fat and considered good food. It was not uncommon to bury them in earth for 24 hours to make them softer and taste less fishy. Soup made from them was said to resemble hare soup. He noted the presence of a huge bank of shells, cast up by the sea some 20 years earlier. He thought that this could be used as fertiliser but the farmers thought otherwise. At Kerbister he found millfoil (*Achillea millefolium*) being dried for medicinal purposes. It had acquired the corrupted name of 'mean-and-folie'. The flowering tips were infused like tea and the extract drunk to dispel melancholy.

On August 3rd he reached Sanday, where there was fine pasture. Free of peat, the sward was dotted with plants of gentianella and bird's eye primrose. Sanday was regarded as the granary of Orkney. Nowhere else had he seen better fields of oats, all of the black sort. Off shore there was a sandy bay occupied by one long cockle bed. In times of dearth dozens of people could be seen at low tide digging for cockles. But this was meagre fare and he remarked again on the irony that the surrounding sea was full of fish there for the taking. He thought it would be a great boon if every Orkney island were provided with one or two fishing families, with proper nets and tackle and hence able to ensure a regular supply of food. There was an unfortunate practice here of paring the sward for use as fuel, since there was no peat on the island – although it could be got on the neighbouring island of Eday.

He drew attention to the flat, uninhabited island of Auskerry, which was endowed with such a rich supply of kelp that this desolate spot probably provided more profit to its owner than a farm of several hundred acres in the best parts of the islands. Neill was particularly interested in sea-weeds. He contributed the article on 'fuci' for Brewster's *Encyclopaedia* (1830).[6] The term 'kelp' included all the larger seaweeds like the yellow, black and prickly tang as well as several species growing in deeper water, including laminarias, which were dragged up at low tide with the aid of hooks. Where the seaweed grew on large, flat rocks, it was unwise to harvest it too close to the rock since if scraped clean, barnacles would invade the site and inhibit more weed from growing there. When the seaweed was fairly dry it was burnt in a kelp furnace, i.e. a hole in the ground. When the hole was nearly full of burnt weed it was raked over, having by then acquired a shiny, gelatinous consistency. It was then removed and stored in a warehouse for export and use in the manufacture of soap and other items. He estimated that between 2000 and 3000 tons of kelp were annually harvested in Orkney. The price for it at Leith, Newcastle and other places varied between £7 and £9 per ton of 21 hundredweight. In Orkney every consideration was sacrificed for the kelp industry, so much so that according to his information, less grain was being produced than thirty years previously. If the market for kelp collapsed – as it later did – the islanders would regret their neglect of agriculture. He reiterated his concern that the fisheries were not adequately exploited. He observed many whales offshore and many gannets diving – evidence of plenty of fish – and yet there was little attempt to catch them. Although cod and haddock might be caught there was a lack of salt to cure them.

From Sanday he crossed to Eday which was mostly marshy heath. Poor quality salt was produced here, although the tax was so heavy that people tried to avoid paying it. The cottagers produced salt on a small scale and bartered it for meal and the like. From here he crossed the sound to Rapness in Westray, where he savoured the favourite fish supper of fillocks (piltocks) which were so abundant. Here he saw a fine field of flax, although there was good deal of the weed gold of pleasure (*Camelina sativa*) growing in it. He inspected the ruins of the Castle of Noltland.

From Westray he went to Papa Westray – 'a beautiful little island' – where there was good soil that supported a rich crop of both red and white clover. In former times good crops of oats, bere and potatoes had been grown. On the margin of the lochs he saw phalaropes, which were known locally – and quite appropriately – as 'half-webs'. He thought the island would be well placed for white fishing, for there was a good cod bank only about two miles off Moulhead of Westray, but no one fished it. However, every cottager fished for his family but never strayed far from the shore. There was a single owner that lived amicably among his tenants all of whom appeared content with their lot. Neill thought them possessed of a kind of languor engendered by their isolated and tranquil existence. Nevertheless, Papa Westray worked its charm on him and he left the island with regret.

On August 9th he set off for Rousay. He thought the boatmen incompetent and avaricious, a consequence of the lack of any regulation of the ferries in Orkney. If the boatmen reckoned they were dealing with a stranger they would increase the charge, proffering all manner of specious excuses for the high price. Anyway, by the time he got to Rousay he was drenched and had to retire to the manse. Observant as ever, he noted the great quantity of the bottle sedge (*Carex rostrata*), species of aquatic bur-reed (*Sparganium*), bulrushes and Yorkshire fog grass. These were gathered and dried to produce what was optimistically termed 'meadow hay', which only the hungry beasts of Orkney would accept as fodder. Although some of the sweetest species of grass grew wild in Orkney there was no attempt to cultivate them. This was a heathery island, with three species – *Erica tetralix*, *E. cinerea* and *Calluna vulgaris* – everywhere, providing food for the abundant grouse. Among the plants noted in one small area were honeysuckle, the soft-downy rose (*Rosa mollis*), brambles, rose-bay willow herb, strawberry, angelica and what he called the great wild valerian (presumably *Valeriana officinalis*), an unexpected find for Orkney.

On August 11th he left for Aikerness on the Orkney mainland and from thence to the island of Hoy, which he reached on August 15th. Here he went up the Ward Hill, admired the magnificent view after the mist had cleared, and was pleased to find *Lycopodium annotinum* and three other species of clubmoss. There were plenty of blaeberries which people gathered and took to Stromness. He did not think much of the Dwarfie Stone but thought Hoy the most interesting place in Orkney for the naturalist. Juniper was common and there were also a few stunted rowans. He noted pigs grubbing up something on the hillside. This turned out to be roots of the tormentil which were used for tanning by the local people. In summer, short eared owls were common and offshore there were plenty of gannets, taminories (puffins) and shelduck – known locally as sly-goose from their habit of enticing away anyone who approached their nest. He particularly admired the township of Rackwick, sheltered by massive cliffs and looking out on a fine bay.

He had now finished his brief tour of Orkney, but before describing his Shetland trip, he made a few general observations about the islands. He found Orkney treeless and did not encounter a single specimen of either pine or birch – although this had not been formerly so since the remains of both species were often dug up in peat mosses. In the parish of Deer, bushels of hazel nuts were occasionally dug up, black and firm in texture. Rowan trees grew in Hoy, together with several species of willow, although these seldom exceeded four to five feet in height. He thought it would be a good plan to import young pine trees from Norway – an equally northern country where they grew abundantly.

Sheep were not rated highly in Orkney, where in many places the wool was still plucked by hand. He came across no honey bees, although he thought there was plenty to sustain them here. Hares were unknown, although rabbits were common. It was a great place for the wild fowler. Apart from grouse on some of the islands, three sorts of geese – the barnacle, the black and the brent – and several kinds of duck – such as wigeon, teal, mallard and golden-eye – provided good shooting. He remarked that two sorts of large, exotic nuts, called Molucca beans, were often washed up on the shore. He thought they came from the Caribbean.

He concluded the description of his Orkney visit with some specific criticisms of the economy of the Islands, very much from the standpoint of a mainland improver. A substantial proportion of the Orkney farms were small and rather poor. They rented at £5–£10 per year. The islanders' boats were only miserable cobbles of one to two tons, since they could afford neither adequate equipment nor more serviceable boats of five to six tons. He hoped the Highland Society for the Improvement of Fisheries would help. He listed at some length the principal obstacles to improvement: the small size of the farms and the lack of leases (or their shortness when granted); the lack of enclosures (and hence the great extent of open, unimproved common land); the total neglect of herding – the livestock wandered at will about the land for half the year; the lack of markets where the fisherman or small farmer could sell to the highest bidder; the need to improve the quality of the kelp manufacture; the ignorance of the islanders as how best to develop the fisheries; the almost total absence of roads, the heavy duty of various kinds paid to Lord Dundas as donatory of the Crown and tacksman of the bishopric of Orkney; and the great intermixture of property in many places. He also drew attention to the total lack of any means of defence. Even in Cromwell's day there had been an armed fort.

On August 24th he was offered passage to Shetland on board an armed vessel that was lying in the Kirkwall Roads, and reached his destination two days later, anchored in Bressay Sound. He had scarcely landed when the local people enquired whether he had

View of Lerwick and Fort Charlotte. The Scots Magazine, Vol 67, 1805

come from Scotland, as if that were a foreign country. At Lerwick, he found that the Dutch and Danish coins in circulation were more common than British coinage. A stuer or stiver (a small piece of base metal with a silvered surface) was worth a penny. The Danish six skilling was worth five pence and so on. Lerwick consisted of one principal street next to the quay, about an English mile long with lanes branching off it. He thought there were about 1000 inhabitants. In the town he found there was a straw-making factory run by a London company, which employed more than 50 women who were at work in two rooms. They were paid one penny a yard and could produce 12, 16 or even as many as 20 yards in a day. Before this factory had existed there had been no such work for women in Lerwick. The nearby Fort Charlotte – named after the Queen – protected the town with the aid of several large cannon which covered the harbour. It was originally established during Cromwell's time and completely restored in 1781. At the time of his visit it was garrisoned by part of the 6th Royal Garrison Battalion. The following day he learnt that several hundred Dutch busses regularly anchored in Bressay Sound in early June, ready for the herring season.

On August 28th he left for Unst in a large open boat. He first landed at Gossaburgh on Yell, where he found the fields so small that they resembled garden plots. In place of the plough, a kind of awkward spade was used by the men to till the ground while the women and children had to drag the harrow. Only the black oat was grown, mixed with tall oat grass. The white oat of mainland Scotland was hardly known. Bere was also quite extensively grown. However, he saw some promising lazy-beds of potatoes, although he thought them planted too closely.

His next stopping point was Ulyea Sound in Unst, where he remarked on the varied nature of the rocks. Beneath the gables of the houses there were lines of small coalfish, called piltocks, strung up to dry without salt. Here he found growing in the kitchen gardens a variety of artichoke with very small heads, scarcely larger than the common spear thistle, but preferred to the larger sort. Neill was surprised and shocked to find no parish school in Unst, despite of a population of about 2000 persons, many of whom did not know the alphabet. He learned that 30 years previous to his visit, several of the inhabitants still spoke Norns – the old Norse language which was now quite extinct.

From Unst he turned south and went by Yell to Bressay which he crossed on foot to visit Noss Island and see the famous cradle, not before being frustrated by the boatmen pulling in opposite directions. A rope system had been set up to transport a man and sheep to the top of a stack some 160-200 feet high and more than 100 feet offshore, as described by Martin in 1703.[7] The top of the stack was covered in fine turf that suited the sheep very well. He understood that the tenant of Noss paid a rent of £50 a year for the whole island and was free to fish for tusk and ling. This was an advantage compared with the usual situation where the tenant was generally bound to deliver the fish he caught to the landlord at a stipulated rate and hence barred from selling it in the most profitable market. The milk and butter produced on Noss were first class, but the owner complained that he could not sell it in the Edinburgh market on account of a Lowland prejudice against Shetland butter.

The following day – September 5th – he was on Scalloway, which presented only barren hills and marshes devoid of botanical interest. He passed soldiers from Fort Charlotte fishing for trout in the loch. He observed that in Shetland, bread only appeared in the houses of the wealthy and potatoes of good quality served as substitute. He noted the Norwegian term 'lax' meaning salmon, in names like Laxforth or Lexforth. In the western Mainland he visited the ruins of Scalloway Castle, built for the detested Patrick Stewart. Not far away, he noted the small green island in a shallow loch at Tingwall. He returned to Lerwick from Scalloway by the only properly made road in Shetland, which owed its existence to the enterprise of two local landowners.

By September 6th it was getting too late in the season to travel further since the weather was likely to become disagreeable and stormy. Indeed so, for he was detained two days in Orkney by a dreadful gale from the south west. After the wind had died down there was a tremendous swell which produced a tumbling motion of the vessel, 'very apt to occasion nausea'. In the evening of the second day after leaving Orkney the vessel passed the May Light at the mouth of the Frith of Forth, and they got sight of the bright beams from the new lighthouse on Inch Keith.

As in the case of his Orkney visit, he concluded with some general observations about Shetland. The inhabitants, although poor, were addicted to tea. Happening to enter a miserable bothy a couple of miles from Lerwick, he found the earthenware teapot simmering. Their regular diet consisted of dried and often putrescent fish and coarse red cabbage. These fish or 'piltocks' were caught off rocks with bait of parboiled limpet, a few of which were stored in the mouth ready for baiting the hooks as quickly as possible. Some of the people had tasted neither bread nor oatmeal for five months. Piltocks were the staple diet, morning, noon and night. Several sorts of shellfish were eaten, including oysters, cockles, mussels, razor shells as well as species of *Tellina* and *Mya*, which were much esteemed.

There were no grouse. Trees were only to be found as remains in peat mosses. There were no lighthouses, although at least three were sorely needed. Communication with the outside was poor and haphazard. Sometimes letters posted from mainland Scotland two or three months apart would arrive at the same time. There were no Justices of the Peace and no Magistrate other than the sheriff-substitute. None of the freeholders ever qualified to vote. The ordinary people appeared to be in a state of vassalage, which was not solely due to the tyrannical spirit of their landlords but partly to the natural poverty of the country and the legacy of the former Danish customs and regulations. Tenants looked to their landlords from a state of hopeless poverty and dependence. They seldom had leases of more than two or three years and were often deep in debt. He thought the landlords preferred to keep their tenants dependent. For every lad that went to the Greenland fishing in summer, there was a levy of one guinea to the landlord – if it were not paid, it would be added to the annual rent and this was generally accepted. The value of Shetland estates depended not so much on the value of rents but rather on the fishing which tenants were obliged to follow. According to the Statistical Report for Dunrossness, the rents were principally paid out of what was caught at sea. Tenants were paid by the landlord 3d per ling and 1d for a cod or tusker (1792). These, when salted and dried, would fetch four or five times as much in the

Hamburg market. Added to the profit from sale of goods such as tobacco, spirits, hooks and linen sold to the islanders at double or triple the purchase price, even allowing for expenses the landlord's profit was very great. The system was open to abuse. Pressed to buy such goods, the tenant could find himself in debt. The landlord provided a small boat and fishing gear to enable the cottars to fish 30–40 miles out to sea. In bad seasons, when the crop failed, the tenants were entirely dependent on him. Were the State to intervene to rescue the tenants from such wretched dependence, it would be necessary to recompense the landlord. Many of the rents had not been raised during the preceding century, indicating how well the landlords were doing out of the fish their tenants caught.

To rectify some of these disadvantages, Neill advocated the establishment of fishing villages with warehouses and a market, where the fishermen could both sell their catch and purchase equipment. The landlords should be encouraged to offer longer leases so that farming – as well as fishing – could benefit. It would also be an advantage to replace the payment of debts in terms of oil, butter and wool, by payment in sterling. He concluded by acknowledging that various pamphlets had already been published on these topics but since they were out of print, he was unaware of their contents. He hoped that the rectitude of his intentions and the general and impartial nature of his observations would serve as sufficient apology for any occasional infelicity of expression.

When Neill's criticisms of the relations between Shetland landlord and tenant appeared in *The Scots Magazine*, they evoked fierce rejection by a correspondent who wrote under the pseudonym of 'Thule'. He referred to similar criticisms by an earlier writer, equally resented as a serious attack on the integrity of the landlords. These criticisms had been rebutted in a pamphlet published on their behalf, as well as in a letter to the Highland Society. A further attack on the landlords had followed some time later, under the name of 'Vindicator', and this had also been rebutted. There the matter had rested until Neill's article appeared. Thule hastened to accuse him of simply repeating Vindicator's misrepresentation. As well as rejecting Neill's general criticism of the dependence of the tenants on the landlords, the writer sought to show his ignorance of the subject by quoting apparent errors in his account. Thus Neill had expressed surprise at the lack of a school in Unst and that the common mouse did not occur on the island. Neill had referred to the castle of Scalloway as square, whereas it was really eight sided with none of the sides of the same length. There were further criticisms, even including Neill's alleged use of Scotticisms rather than standard English, which was rather like the kettle calling the pot black. Thule complained to Constable for publishing such material, who rebutted his criticisms in a rather acrimonious correspondence. However, Thule had picked the wrong adversary, for Neill did not budge and counter attacked. He had neither read nor been aware of the earlier writings about Shetland. Unst had no established parochial school, whatever Thule might assert to the contrary. As for the mice on Unst he had merely quoted the Statistical Account of Scotland written by the parish minister, although he later conceded that the absence of mice applied to the island of Uyea. In describing Scalloway Castle, Neill had used the term 'square' as an adjective rather than substantive so this disagreement was but a quibble. When it came to the condition of the tenantry, Neill became more caustic. Thule had stated that the poorest

Shetland tenants were more independent than substantial tenants in the south. Neill replied that this landlord was pleased to argue that the tenants' 'independence' lay in their poverty and that being fishers, they could become sailors in a moment. But this was a miserable compensation for a poor Shetland tenant with a large family. Thule was expressing 'more of the cloven hoof than his brethren will probably thank him for'.

In further exchanges of letters Neill rejected the claim that he was an associate of Vindicator, merely repeating earlier criticisms. On the contrary, he had no idea who Vindicator was and had seen none of the earlier criticisms. He wrote: 'I would be sorry, after all to accuse the anonymous Zetlander of intentional falsehood; but I at least affirm, that he has fallen into the greatest mistake, and has indulged in personally injurious insinuations with reprehensible carelessness. While he declines to undertake the responsibility, which would attach to his name and character, I feel myself, in this instance, called upon to follow a different line of conduct.' In another jibe his critic had accused him of 'bawling about oppression' and objected to his use of the term 'vassalage', at which Neill quoted Dr. Johnson to justify its use. This letter concluded: 'I assure the champion of the Shetland landlords, that my remarks had no object in view but the unimpeachable one of improving the situation of the natives. Conscious of upright intentions, I shall never be influenced by his unnecessary invective, timidly to abandon my statements or opinions, and I would remind him that to declaim and vilify, is a different thing from disproving or refuting.'

Given that other independent observers had commented on the exploitation of the poor peasantry by the Shetland landlords, we can infer that Neill had a valid case, notwithstanding the anger of the landlords' spokesman. He had the last word, since the preface to his book was wholly uncompromising. He repeated his view that the greater part of the Shetland tenants appeared to be sunk in a state of the most abject poverty and misery. They were even without bread, living on a diet of fish and cabbage, often sharing the same roof with their cattle. Their little pieces of land were neglected since they were obliged to spend the summer at the ling and tusk fishing. Neill had no doubt that the tenants' ills stemmed from the obligations owed by the tenants to the landlords.

There was a final twist to the story. Neill had discovered that Vindicator was a respectable clergyman living in the Edinburgh Canongate. Neill had asked the clergyman whether he wished to make any further comment and he replied: 'I have no wish to reply to malignant, ill written nonsense. I would invite the sagacious "Friend of Zetland" [the latest nom de plume of the landlords' protagonist] before he attempt publishing again, to learn to spell, to write grammar, to show common sense and have good manners.'

Apart from the polemical exchanges between Neill and the spokesman for the Shetland landlords, the appendix contained several articles of scientific interest. Thus Dr. T. S. Traill contributed observations on the mineralogy of the Shetland Isles obtained during his visit in 1803. Sir Alexander Seton of Preston wrote about the economy of Shetland and how it might be improved. Neill contributed a substantial list of plants native to Orkney, omitted in earlier accounts of the flora. He added more than a hundred angiosperms, 34 species of moss, plus sundry corrections. There was also an interesting article by Neill on the local names for seabirds. In addition to the names already mentioned, the following are worth noting: bawkie for razorbill; ratch for little auk; allamotti for stormy petrel; malmock for fulmar; lyre or lyrot for shearwater; scarf for shag; raingoose for red throated diver; loon for speckled diver; toist, tyst or taistie for black guillemot; swart for great black backed gull; and syth for coot.

Neill's account of his northern trip reveals him as a keen observer of both the natural and human scene. In the spirit of the times, when agricultural improvement was uppermost in many minds, he was concerned to discover ways in which the islanders' lot might be improved. Once he came to an opinion he stuck to it and was both ready and very able to defend it. His description of the trip and the accompanying verbal fisticuffs with the Shetland landlord created considerable interest. A reviewer of his book praised his public spirit and regretted the obloquy he had incurred. Neill had established a reputation as an informed commentator whose opinions were to be treated with respect.

1   Patrick Neill, *A Tour Through Some of the Islands of Orkney and Shetland, with a View Chiefly to Objects of Natural History and also Occasional Remarks on the State of the Inhabitants, their Husbandry and Fisheries,* 1806, Constable & Co., Edinburgh.
2   D.25/1/18 Letter from an editor to 'Thule', care of Thomas Mouat, March 25 1806, declining a contribution anent Patrick Neill, Shetland Archives.
3   D.25/119 Letter by 'Thule' to Messrs Constable & Co., 7th April n.d., concerning a contribution anent Patrick Neill, Shetland Archives.
4   D. 25/1/20. Letter by 'The Editor' to 'Thule', February 8 1806, asking for a final contribution concerning Patrick Neill, Shetland Archives.
5   D. 25/ 1/21. Letter by 'The Editor' to 'Thule', February 24 1806, Shetland Archives.
6   Patrick Neill, *Fuci, The Edinburgh Encyclopaedia, Conducted by David Brewster with the Assistance of Gentlemen Eminent in Science and Literature in Eighteen Volumes,* 1830, Vol. X, pp. 1–23.
7   Martin Martin, *A Description of the Western Isles of Scotland circa 1695,* 1994, p.379, Birlinn Ltd., Edinburgh.

4 🌿 Nature Notes and Observations

Although field botany and the cultivation of choice plants comprised an abiding passion throughout Neill's life, there was no aspect of the natural scene which failed to captivate his curiosity and interest. Although his observations did not appear in book form, his extensive contributions to *The Scots Magazine*, the occasional paper in the Proceedings of the Wernerian Natural History Society and his correspondence with naturalist friends, leave no doubt about his fascination with the ways of animals and plants, the geology of the landscape, the vagaries of the weather or indeed any natural phenomenon likely to provoke comment or speculation. Nowhere is this better demonstrated than in his monthly 'Memoranda on Natural History', contributed for a decade from January 1808 (Vol. 70) to June 1817 (Vol. 79) to *The Scots Magazine*, as well as sundry reports published in sometimes unexpected places. Following on from the recognition he had won from the publication of his *Tour of the Northern Isles*, the Memoranda served to consolidate his position as an informed commentator. They provided a running commentary on any event or observation in the neighbourhood of Edinburgh likely to interest a reader with at least a passing interest in the natural scene. But at the same time, they provided him with an opportunity to draw attention to causes dear to his heart or castigate official shortcomings. His more interesting observations are set down here to bring to mind long-forgotten events and social attitudes of the Edinburgh of his times.

Weather

These days, with the threat of climate change in mind, what happened to the weather in earlier times has a compelling fascination. Neill's accounts do not disappoint on that score. Apart from the stock-in-trade of every writer of nature notes, such as the early or late appearance of spring, or when the first cuckoo was heard on April 29 1810 at Drumsheugh, it is the more dramatic manifestations of weather that engage our attention. Thus in February 1808 Duddingston Loch and Loch-end, frozen rock hard for a week were covered with skaters and curlers. It was equally cold in England. The London mail, due in Edinburgh on February 14th was delayed by snow and wind and did not arrive until three days later. By contrast, there was a heat-wave in July of the same year. In England several people died of sunstroke and exhaustion while many lambs and wild birds perished from the heat. There was a water shortage. Butter sent to London arrived as oil. Later that year in the first half of September, excessive rain damaged the wheat crop, which was attacked by mildew, rust and blight. On October 14th there was a heavy snow shower. On the same date a meteor flashed across the night sky. By mid-December the skaters were back on the lochs.

The following year, the first week of May was intensely cold, followed by a thaw and at the end of the month a heavy fall of snow was followed by intense rain. In 1811 the winter was very open and mild. Consequently the stalls of the poultry market were not loaded with mallard, widgeon and scaup duck as in a severe season. In March however, intense frost and gales checked the growth of vegetation, while in May a hailstorm broke greenhouses in southwest Scotland. In June, lightening hit a house in Fountainbridge Street, setting a maid servant's clothes alight so that she was scorched and had to be taken to the infirmary. A former seaman was struck at Craigleith Quay, losing both sight and mobility on one side of his body. In October, although mild, the waters of the German Ocean – i.e. the North Sea – rose some 20 feet due to the east wind and change in pressure. It flooded houses at Methil, East Wemys and Dysart, sweeping one of them away. At Kirkcaldy the sea-dykes were breached. Similar damage was reported all along the east coat. At Stonehaven the tide damaged ships in the harbour and several streets were under water.

In 1812 a foot of snow fell in March, with 15–20 feet drifts about Arthur's Seat. On March 22nd the London mail reached Edinburgh but had to leave by horseback since the roads were impassable for coaches. On March 27th ponds were frozen and it seemed odd to see boys skating on North Loch ditches on Good Friday. Even by mid-April the Pentland Hills and Arthur's Seat were streaked with snow.

By contrast in 1813, February and March were fine and mild. By February 10th snowdrops were in flower in the garden followed by winter aconites and hepaticas by February 16th. But in 1814 it was a very different story, for there was an intense frost in January, worse than that experienced in 1793. Five degrees below zero was registered in Glasgow. Trees were damaged; hardwoods like elms burst apart with a noise like cannon shot. The Forth was covered with ice floes from Queensferry upwards, with vapour rising between the ice blocks, similar to what had seen in Hudson Bay and Greenland. The water froze in Leith Harbour so that sailors could cross from ship to ship on the ice. The Thames froze and enterprising printers set up presses on the ice to run off sheets commemorating the great frost of 1814. Many perennial plants were killed. Shrubs like Portugese laurel, sweet bay and laurustinus were killed or damaged. In places about Edinburgh the whin bushes appeared dead, overwhelmed by weight of snow.

The following year also saw a severe frost and snow at the end of January, but February was mild, so that chaffinches and goldcrests appeared about Edinburgh in number. But in 1816 there was another severe winter – what Neill referred to as the worst in living memory, with frost, snow and gales which began early in November and lasted until late February. Since mid-November a flock of snow buntings had frequented Calton Hill – a favourite place for bird-watchers, where visitors fed the buntings with barley.

The hard winter was followed by an uncongenial summer. Flowering and vegetative growth was five weeks later than usual. There were no strawberries in July. The bad season was general throughout Europe. In 1817 both summer and autumn were dismally wet. Although winter came early, the weather became mild in January so that by the end of the month, winter aconites were in flower and snowdrops were appearing.

Although variable as ever, the severity of the Edinburgh winters in the early years of the 19th Century often surpassed present day experience.

Birds and Fishes

As an enthusiastic naturalist, Neill was on the alert to report anything of note in the animal world. There are a number of comments about birds seen – or more likely shot, since anyone with a gun seemed impelled to shoot any bird which appeared unusual. His regular inspection of the poultry stalls often turned up something worth noting. Thus in 1808 a couple of guillemots and a red-throated diver entangled in fishing nets were offered for sale. In February a goosander, a couple of mergansers, together with a few woodcock, ptarmigan and plovers were on the stalls. The Newhaven fishermen referred to the Great Northern diver as the 'herdman of the seas'. In the winter of 1811, two specimens of what he called the black-eyed gull were taken at the Bell Rock Floating Light and sent to the University Museum. In February of the same year a little bittern the size of a fieldfare was shot at Tyningham by the Earl of Haddington's gamekeeper. Neill recorded this as the first occurrence of this species in Scotland. In 1812 a water rail was shot in a meadow at Comely Bank. In 1812 a hoopoe had been seen at Stronsay, Orkney. In 1814 he referred to a sailor who amused himself with his gun and had probably killed a greater number of rare birds than anyone else in the district of Cramond. A sea swallow or pictarney of unusual appearance that had attracted his aim turned out to be a roseate tern.

In 1814 Neill had a comment about the great auk of which there had been no record from the Northern Isles during the preceding half century. A Mr Bullock from London, when visiting Orkney, had been told that the 'king and queen of the auks' frequented Papa Westray in summer. A female was killed with a stone while incubating her eggs, although the male escaped. In the summer of 1813 a solitary male was seen in its previous haunts, but the islanders killed it and sent it to the museum. Neill claimed this was the only specimen known to exist.

In 1814 an osprey was seen fishing on the loch at the Hirsel. Lord Home succeeded after several days in shooting it – the usual story. This also was sent to the University Museum. The same year Neill noted that a large shrike or butcher bird was shot at Bavelaw by the gamekeeper. In late 1816 a bittern was shot at Dunbar. Thirty years previously the bittern was not uncommon in the marshes of the Lammermuirs, but was now so scarce that the sighting of a single specimen was news.

Neill had a particular interest in the sea and all things marine. He regularly visited the fish market and kept in touch with the Newhaven fishermen who would let him know about anything unusual. He often referred to the 'inexhaustible' supplies of herring in the Firth of Forth or 'Frith' as he always called it. In 1810 he noted there were more than 100 boats fishing between Queensferry and Cramond. Cuttle fish were often cast up on the shores of the Forth. They were known to the fishermen as o-fish or sleeve fish. Sprats – known locally as 'garvies' – were taken in large numbers but seldom appeared in the Edinburgh market. He thought the local ones were larger than those caught in the south of England.

In 1808 he reported that several open boats from the Moray Firth had appeared in the Forth with cargoes of dried fish, especially cod, ling, haddock and dogfish. However they were dried without salt and were not well cured. He thought this was the first time dogfish had been offered on the Edinburgh market. The Buchan fishers obtained oil from the dogfish livers. For drying, the fish were split down the back, the guts removed and then spread on rocks. Either boiled or broiled they tasted oily and rancid. Its name did this fish no favours, prompting a Mr Knox – who had written about fisheries – to suggest renaming it the King George fish. Needless to say, nothing came of this marketing ruse. Later the same year, a society of gentlemen had brought a band of fishermen from England with a decked smack, to fish for sole and halibut in the Forth. They had little success and they soon left. When naval vessels were anchored in Aberlady Bay the sailors passed the time fishing for sole, with enough success to supply their mess and local houses. In August 1810 Neill again commented on the vast numbers of herring in St Andrew's Bay and the mouth of the Forth. At Dunbar, boats were taking 5000–6000 in a morning. There were also very large shoals of Piked Dogfish which damaged the herring nets.

In 1811 Neill reported the capture of a swordfish near Queensferry; the third time this had happened. It was sent to the University Museum but not before it was sampled and reported to taste no better than a large halibut. September was the haddock season. About 30 years previously, haddock entered the Forth in immense numbers. About 1783 it seems they disappeared, but came back in a smaller number in 1795. The reason for such behaviour was unknown, but Neill speculated that storms might have moved mussel beds or the herring fry on which the fish fed, so they had gone elsewhere. In 1812 a sunfish had been taken at Newhaven. In the following year, two rare fish of the genus *Trichiuris* – hair-tail – had been taken at Port Seton in Aberdeenshire. The local people were so surprised at finding these eel-like fish that they conveyed the bigger one to Gordon Castle for inspection by the Duke. The fish was cooked and pronounced tasty. Mr James Hoy, of the castle, sent a description to the Linnaean Society.

In 1815 several shad were caught in the estuary of the Tay. Although common in Europe, it was not recognised as edible in Edinburgh. In 1816 Neill wrote a long article on fish ponds, especially the ones he had seen in a sea inlet in the Rinns of Galloway where the fish were kept for family use. There was quite a collection of species which included cod, haddock, coalfish, whiting, pollack, salmon, and also flat fish. The keeper caught these by handline and transferred them to the ponds. Sand eels together with limpets, mussels and herring cut into small pieces were provided as food.

Whales

The appearance of whales always merited comment in his memoranda. In 1808 a bottle-nosed whale had been stranded on the East Lothian coast and was immediately stripped of its blubber by the local people. In the autumn of the same year a large whale had been stranded at Alloa. Neill went out to examine it but was forestalled because the locals had already removed the blubber, for which they received £15. This was a baleen species that he identified as a pike-headed whale. This specimen measured 43 feet in length.

In 1810 a school of about 500 whales had appeared in Orkney at Stronsay, where they had been driven ashore. The largest was about 22 feet long and there were also many young of five to six feet. These were known as 'ca'ing whales' in Orkney, sometimes wrongly named bottle-nosed whales, which was a different species. The same year a number of small whales – apparently killer whales – passed Queensferry, heading up the Forth. At Alloa the local distiller organised his workmen to pursue the whales in boats and attack them with guns, spears and knives. Some of the survivors were killed further upstream at Tullibody and another was pursued and killed off Stirling. In all 19 animals were slaughtered. In the autumn of 1815 a beluga whale was seen swimming up and down the Forth. As it floundered in the shoals off Cambuskenneth Abbey it was despatched by the locals with muskets and spears. The 13–14-foot carcase was presented to the University Museum. Evidently the sight of a whale offshore galvanised people into an unrestrained frenzy to kill it. What a contrast with present day attitudes!

A sea monster

In 1808 Neill became very excited by the report of what he described as a 60-foot sea snake that had been found on the shore in Orkney, although it was in a damaged state. Indeed he became so excited that he read a paper to the Wernerian Society, describing what he had heard about it. He suggested that it was none other than the sea-ormen or *Serpens marinus magnus*, discussed long ago by Pontoppidan in his *Natural History of Norway*. He even proposed the names *Halsydrus* for the generic and *Pontoppidani* for the specific name. Later he had to concede ruefully that this was premature, since the rather battered remains appeared more likely to have been those of a shark. There is an intriguing sequel to this story. The remains were kept in the collection of Dr. Barclay, a local anatomist. When he died his collection was dispersed and, in a letter to the geologist Charles Lyell, Neill thought the remains must have been thrown out.[1] However, in a recent report in *The Scotsman* newspaper (September 5 2008) it was stated that they still exist and are to be examined for DNA evidence of their origin.

While in euphoric mood at having apparently found a rare sea monster, Neill was even willing to consider the question of mermaids. He referred to a recent report from Caithness in which two young ladies had observed a creature they identified as a mermaid. They had watched it for about an hour and were able to distinguish it from a seal. He wondered whether they might have seen an angel fish, although their minute description of the hair on the head baffled conjecture. He thought the witnesses were to be commended for their courage in describing what they thought they had seen, in the face of likely ridicule. He was all for keeping an open mind as to what strange animals might lurk in the ocean.

Plants

Neill was a botanist and gardener with a life-long interest in horticulture. He kept an eye on what was flowering in the Botanic Garden and in local nurseries, especially Dickson's, which was the largest. The plight of the Botanic Garden was a recurring theme. The Garden was chronically short of funds and the greenhouse and hothouse were in disrepair, threatening the survival of their contents in severe weather. In 1808 he complained that a fine old trailing fig planted 30 years previously, had succumbed during the winter. A handsome camphor tree had to be cut back to prevent its crown hitting the glass roof. Surely such neglect could not be known to the Government. Two years later Neill reported that the superintendent, Thomas Somerville, had died at the age of 27 and that the Garden was looking for a successor. But the salary on offer was but a paltry £40 per year. An experienced person could get £60 or £100 in London. What was needed was a subvention of £600–£800 to provide an adequate salary for the superintendent and the wages of several gardeners. However, Neill could later report that a successor had been found in William McNab, who gave up his better-paid job at Kew to come to Edinburgh. His appointment proved to be a lucky choice, for McNab went on to become the most famous gardener in Scotland and one who made his mark on horticultural history. He was also a close friend of Neill who wrote that at the time of McNab's appointment, he had brought a number of rare plants for the Botanic Garden greenhouse and stove.

In 1812 Neill returned to the financial plight of the Garden. Another plant – a large and elegant date palm – was pushing up against the roof. He could hardly believe that the true state of the garden was known to the Prince Regent and his advisers. In 1813 the predicted fate of the dragon tree had come to pass. It had to be cut back and yet it would only have cost £100 to raise the glass roof and so preserve the finest specimen in Europe. It had reached a height of 29 feet, compared with the 14 feet of the one at Kew. Such neglect was a national disgrace. The following year he reflected that the improved accommodation of the Botanic Garden at Leith Walk had obscured its continuing needs. The location of a new garden was under consideration. Neill hastened to offer advice as to the need for it to be located in Edinburgh, properly sited and provided with good soil. He advised there should be a sylva where gentlemen could become familiar with the different species of tree. There should be a section devoted to vegetables and fruits suited for cultivation in Scotland. There should also be an area set aside for experimental studies in agriculture and horticulture, perhaps an acre at the disposal of the Professor of Agriculture and another under the supervision of the Horticultural Society. This arrangement would be useful for botany students and could even contribute to scientific advancement. This all sounds very much like a blueprint for the Experimental Garden which was later established at Inverleith by the Caledonian Horticultural Society, of which Neill was secretary.

Neill criticised the Government for not supporting the publication of key botanical works. In 1810 he expressed the hope that the Government would publish a full account of Flinder's voyages, to vie with the French 'labours and pretensions' of Baudin and Hamelin. It was whispered that the Government would publish the maps but not Robert Brown's

Prodromus of the Flora of New Holland although Jussieu – even in time of war – recognised Brown as the foremost botanist of his age. By comparison, the poorest government in Europe published the most trifling discoveries made in the Baltic, Iceland, Greenland or the coast of Norway, in elegant volumes. How could the Government resist the recommendations of the President of the Royal Society of London? He compared this parsimony with the outlay of £60,000 for what was called 'the Fete in the Parks', of a trifling, if not ludicrous nature, chiefly devised for 'the wonderment of a mob or, at the best, for masters and misses'. By contrast, a mere £200 to £300 would cover the cost of Robert Brown's *Flora*. This was indeed a 'national stigma'.

Neill was also exercised by the great dearth of books on botany in the Edinburgh libraries. He listed a number of contemporary works which were too expensive for ordinary folk to buy, but which should be available in public libraries. For example, Smith's *Flora Britannica* was lacking, even though Smith did medicine at Edinburgh and first botanised in Ravelston Wood. The lack of such books could hardly be due to lack of funds but until rectified, students of botany must labour under 'intolerable disadvantage'.

Neill was impressed by the recently erected conservatory at Dalkeith, where there was also an excellent collection of American shrubs due to the skills of the head gardener – his friend James Macdonald – and the 'taste and liberality of the noble family of Buccleuch'. Neill always displayed a respectful attitude to the gentry. His monthly reports often referred to the first flowering in Scotland of recently introduced exotic plants, usually in the stove or greenhouse of the Botanic Garden, local nurseries or at Dalkeith. For example in March 1816 he noted that *Azalea indica* – introduced to Britain about 18 years previously – had flowered in the Dalkeith hot-house. Other references were to newly-introduced Australian species, chiefly from Brown's collecting during Flinder's explorations.

Neill was a friend of the maverick botanist George Don who at that time was running a chaotic nursery and plant collection at Doohillock by Loch Forfar. In 1809 Neill noted the number of species in Don's collection as follows: *Veronica* (55), *Salvia* (50), *Campanula* (44), *Allium* (40), *Saxifraga* (46), *Cucubalus* (13) *Fumaria* (14), while for *Ononis*, *Vicia* and *Lathyrus* almost all the known species were represented.[2] He had a particularly good collection of Scottish alpines, many species of *Carex* and dozens of different grasses. He even had *Panax* or ginseng in flower. However his garden was notoriously untidy and bore little relation to the conventional appearance of a garden, by way of gravelled walks and tidy borders. In 1810 he reported that Don had been plant collecting on Ben Lawers and neighbouring hills for the fourth or fifth time and had discovered hitherto undescribed species of *Carex*, *Eriophoprum* and *Cerastium*, however not all of which were subsequently confirmed.

In November 1814 after a prolonged period of drought, he noticed that seeds had unexpectedly ripened on a number of species like the everlasting pea, *Lobelia cardinalis*, *Tradescantia virginica*, *Fuchsia coccinea* etc. He speculated, erroneously, that having produced seed in the open, they would prove hardy in the future.

In 1810 he had a comment on fiorin, more recently known as creeping bent grass (*Agrostis stolonifera*). It had been advocated as a food for cows by an Irish writer. In a recent account by another proponent it was claimed that cows in Ireland were seen grazing on it in a quagmire, where flats were needed to keep the farm workers afloat. There was no word of how deep the cows sunk. Sir John Sinclair – ever on the look-out for potential agricultural improvements – had started growing it in several places around Edinburgh. Neill reported that one such experimental well-irrigated plot at the back of Heriot Row, was now in luxuriant growth.

In 1815 he had quite a long account of a new species of *Urtica* or stinging nettle, recently discovered in Canada about the Great Lakes. He gave a detailed description and suggested that it might prove a substitute for hemp. In former days, a small quantity of hemp was grown by every farmer in Scotland for use on the farm. Hempen ropes used to be made by ploughmen on winter nights. It took the place of leather but now it was scarcely cultivated. He had seen staple made from the Canadian plant and thought it of good quality. He suggested that this new species might do well in the Hebrides.

Another plant deserving wider notice was the sea kale (*Crambe maritima*). Wild plants had long been gathered and eaten in southern England. Neill described how to grow and blanch it for table use. The Caledonian Horticultural Society – ever keen to promote home-grown fare – had offered a prize for whoever brought the largest sample to market but there had been no takers.

Menageries

From time to time, travelling menageries visited Edinburgh and Neill hastened to examine what was on display. In 1808 a Mr Miles turned up with a collection that included the usual big cats, several kangaroos, emus, a spoonbill etc. Neill recommended an inspection but had to mention several errors in the naming of the animals. In 1811 a Mr Polito brought a more representative collection to the city which Neill considered very good. The animals were in excellent condition. There was a male elephant, a hyena, a couple of jackals, a lion and a tiger, to mention the most notable exhibits. When the lion uttered his hollow roar and the male jackal in a nearby cage set up a persistent howl, this strange concert could be heard the entire length of Princes Street. In 1812 a young polar bear about the size of a mastiff was given to the Professor of Natural History. It had been caught as a cub on an ice floe off West Greenland. Neill had to admit that the accommodation for this unfamiliar beast was neither 'so commodious nor so handsome as might be expected from the liberality of the enlightened Magistrates of the Scottish metropolis'. There was no container for water and the roof was too low for exercise. He followed with the observation that the foundations had been laid for a menagerie in Edinburgh and if adequately supported, would result in a good collection – apparently the first stirrings of what would eventually lead to the formation of the Edinburgh Zoological Society of which he was a founder member.

In 1813 the Earl of Morton had procured a year old quagga which was taken to Dalmahoy, where it thrived on the good pasture. However, it fared badly when winter came and died in December. Its carcase was sent to the museum for preservation. Another one was obtained from the Cape and efforts made – without success – to break it in like a horse. The quagga bit its handlers. Both these animals had been male and Neill hoped there might be a chance of obtaining a female which could be mated to the survivor. Unsurprisingly attempts to mate it with a mare proved fruitless. Neill claimed that the name 'quagga' corresponded to the Hottentot version of its cry, which resembled the howl of a large dog. In 1816 Neill reported that reindeer had been imported to Orkney. He had not then been aware that the Duke of Atholl had tried to establish them at Blair, without success. He then went on to extol the virtues and tasty meat of the American wapiti which might do well in England.

Mad Dogs

In 1809 there was a curious story about mad dogs. The Edinburgh magistrates had issued a proclamation that ordered the inhabitants of the city to keep their dogs closely confined for six weeks, since they believed a mad dog was at large. Neill did not think such draconian measures were called for. Just because a mastiff was seen running at full speed through the streets, lolling his tongue and biting any other dogs that got in his way did not prove that the dog was rabid. Given a collar, food and water, such a dog would behave normally; it was probably only a country dog making for the open fields. He did not dwell on the fact that the average person would be unlikely to intercept such a dog to minister to its needs. Although unwilling to criticise the magistrates for their concern for the public, the situation hardly called for advertisements in the newspapers, notices stuck on lamp posts and proclamation 'by tuck of drum', all of which drove people into a state of fear. Besides, there was a sinister outcome. Stray dogs were killed by hatchet wielding police officers but – more insidiously – by apprentice tanners who were always ready to lend their unwanted help on account of the value of the skins that they could thus obtain with impunity. He returned to the theme in March 1809 to report that there had been great slaughter of pet dogs. The precincts of the principal tanneries were strewn with their carcases. There was a need to restrain the tanners' apprentices, who had even been known to entice dogs from their homes in order to slaughter them. All this had taken place without any real evidence of a mad dog.

Lighthouses

Neill had a general interest in lighthouses. He was a friend of Robert Stevenson – the engineer for the Northern Lighthouse Board and Robert Louis Stevenson's grandfather. There is an interesting account of his visit in August 1811 to the Isle of May, in the Forth, where there was a 'light'. The island was part of the Scotstarvit estate, with the right to levy a charge for the maintenance of the light, originally granted by Charles I. He went to examine the 50–60 foot tower that was erected in Cromwell's time in 1656 (although the original had been put up in 1635). He ascended by a 'miserable' stair to the platform, laid with flagstones and with a large stone grate in which a fire was kept burning at night throughout the year. The fire was lit at sunset with the aid of live coals and needed attention every half hour. In gales of wind it had to be fed every 20 minutes. During a long winter's night of strong wind, more than three tons of coal were used, requiring the attention of three attendants. Under such conditions, feeding the fire was a dangerous task. The man adding the coal on the windward side had to hang on tight. For a whole winter's supply, 400 tons of best quality Wemys coal was needed. The fuel was lifted to the top platform with the aid of a clumsy piece of gear.

In 1791 there had been a tragic accident. The keeper, his wife and five or six children had died in their beds. It was supposed that a large bed of cinders – allowed to accumulate over a ten year period – had somehow caught alight and generated the poisonous fumes that had overwhelmed the family. To add to the horror of the story the only survivor was the baby, still attached to its mother's breast. Although at the time of Neill's visit the Isle of May light might have been the best in Britain, it varied in brightness and was really inadequate. Even mariners familiar with the Forth preferred to wait until daylight rather than rely on the Isle of May light. In the preceding January, two fine frigates independently mistook a glowing limekiln near Dunbar for the May light and were totally wrecked.

He took the opportunity of his visit to look around the island and size up the geology. There was little of botanical interest. About 60 sheep were fattened on the island. Near the harbour, on the east side, he found two Newhaven pilots with a yawl, on the lookout for foreign ships. There was a small enclosure where the attendants of the light grew turnips, potatoes and bere. Not long after his visit the ancient light gave way to a fine new lighthouse, built under Stevenson's direction. One odd observation is worth a note. The rabbits of the island included a variety with long, silky hair. These rabbits tended to have their burrows rather apart from the rest. Neill speculated that they might be descendants of a cross with the Angora type.

About a mile and a half off the most easterly point of Fife there was the Car Rock which was a serious hazard for shipping. Four vessels had been stranded on it or lost by collision during the winter of the previous year. The commissioners of Northern Lights had decided to erect a 40 foot conical stone column which Mr Stevenson thought feasible. Neill went to have a look and managed with difficulty to land on the seaweed-covered rock in July. He concluded it was a continuation of a sandstone ridge visible on the nearest mainland. Plans for the column were going ahead. A sandstone quarry had been opened near Pitmilly and an old house belonging to the Earl of Kellie was being fitted up to house the workmen. In preparation for the work, the surface of the rock was cleaned but within six months it was again covered with several species of seaweed. Instead of a light it had been decided to fit a tolling bell on top of the column. Mr Clark – an ingenious Edinburgh clockmaker – had devised a mechanism whereby the rise and fall of the water activated a series of hammers which struck the bell.

In 1811 Neill referred to the completion of the Bell Rock Lighthouse. The supply vessel – which had been moored by mushroom anchors off the rock for just over three and a half years – was brought back to Leith. The bottom of the boat produced a remarkable collection of marine life. Covered with seaweed – mostly *Fucus esculentus* and *F. digitatus* with four to five feet long fronds – there were large mussels, many large acorn barnacles and a host of marine worms. Through the 'obliging attention' of Mr Stevenson, specimens of all these were recovered and presented to Neill.

Miscellaneous

There were a few additional items in Neill's memoranda worth including here. In 1808 a very large rock crystal found at Braemar, was to be seen in the jeweller's shop of Messrs

Marshall in the High Street. It comprised two six sided prisms, weighed 19lb 5oz and cost its owners 40 guineas. There was a considerable demand for such Scotch Topaz.

In September 1809 he noted the hardship of the Highland reapers who had come to Lothian to find that the harvest was three weeks later than usual. Help was provided by local gentlemen and farmers, and also by subscriptions taken at church doors. Mr Nisbet of Direlton had kept a public kitchen for them for a fortnight.

In 1811 a full-grown wolf was found on Tyningham sands. It was inferred that it had been on a vessel which had foundered in the Forth.

In April 1811 grass mown on well irrigated meadow at the foot of Salisbury Crags was being sold in very small bunches at 2d apiece. It consisted chiefly of *Poa trivialis*.

In 1813 he reported on the sailing of Danish vessels with supplies for missionaries and others in West Greenland. This had been interrupted by the recent 'unhappy war' with Denmark. One of these vessels – Hvalfisken – had lately arrived at Leith. Among the passengers were a Mr Geisecke – also the priest of Jacobshavn who had completed his tour of duty – and his Eskimo wife dressed in her native costume. Mr Geisecke, the priest of Jacobshavn, had explored up the coast until he met the pack ice. Three or four years previously he had sent off a collection of minerals on a vessel which was captured by a British cruiser. The cargo and his minerals were sold off. The collection – which included some rare items – was described in the transactions of the Royal Society of Edinburgh and Volume VI of the Wernerian Society. At least Mr Geisecke was glad his collection fell into scientific hands. He hoped to make another collection.

In 1815 Neill had an article on fire-damp. A local chemist – Dr Murray – had developed a safety lamp that had been described before the Royal Society of Edinburgh. Davy of London had also developed a safety lamp. The two systems overlapped in design but Neill preferred Murray's.

In 1815 he described with horror how three of the Standing Stones of Stenness in Orkney had been knocked down by the local farmer. Two of them had been broken in pieces. The third was the famous perforated stone that Neill had described in the record of his Orkney and Shetland trip. This was the stone which played such a part in local custom and tradition, especially the plighting of troth ('the Promise of Odin') when hands were joined through the hole. The farmer had been ordered to re-erect it.

In August 1816 he reported at length the consequences of an earthquake which was felt as a sudden, strong, shock, especially at Croal, 18 miles west of Inverness but also at West Gairloch and Applecross. It was just perceptible in Edinburgh. The shock happened at 11 o'clock at night, accompanied by a rumbling noise and the shaking of houses. He gave a vivid description of the reaction. In the worst affected town almost every inhabitant rushed into the streets – women and children screaming and a very considerable proportion of them naked. Many flew to the fields and stayed there most of the night. Chimney pots were thrown down, a hotel was split from top to bottom and the church steeple was damaged. But despite all the fright and uproar no one was hurt.

In 1817 there was an account of a convenient pocket barometer – a 'sympiesometer' – made by Mr Adie, an Edinburgh watchmaker. A Mr Christie – captain of a Scotch Indiaman

– had taken it with him on his voyage and concluded it was as good as the best marine barometer. This would be a great boon to the geologist and the 'travelling philosopher'.

In the last year or two of his memoranda the reader has the impression that Neill may have been tiring of the responsibility of finding topical news to write about every month. His contributions were increasingly devoted to long didactic essays on such subjects as the properties of mushrooms and truffles and other general topics. Perhaps that is why, when *The Scots Magazine* was reorganised in 1818 with a different arrangement of articles and reports (although learned accounts of natural history and other scientific topics regularly appeared), Neill's memoranda were not continued.

Miscellaneous Observations

In addition to his regular reports Neill had comments and observations on natural history published in journals or books primarily devoted to other topics. Evidence of his standing as the local expert appeared in a contribution to Stark's *Pictures of Edinburgh Containing a Description of the City and its Environs*.[3] This very popular work went through six editions. It contained articles on the natural history of Edinburgh and its food markets. Although the articles were unsigned and attributed to 'an eminent naturalist', there is no doubt from their style and content that Neill wrote them. They are of interest for the inclusion of the Scottish names of many of the animals he noted and also because his account of the markets supplements that of the natural history by way of the species – especially of poultry and fish – which were for sale.

His list of mammals includes the hedgehog or 'hurchin' and the weasel or 'whitret'. He noted that while the Norway rat was too common, the black rat still inhabited the garrets of the tall houses in the old town. Among the more notable birds, the kingfisher was to be found along the Water of Leith, the blue backed shrike frequented Arthur's Seat and the kestrel was to seen breeding on the rocks of the Castle, facing Princes Street. In summer the nightjar – or what he termed 'the goatsucker' – and the corncrake appeared, while in May the cuckoo or 'gouk' was to be heard about Duddingston Loch. Regular migrants included the fieldfare or 'feltefer', the redwing and sometimes the 'bohemian chatterer' or waxwing. Sea birds along the Forth included terns or 'pictarnys', the kittywake, the shag or 'scart', the guillemot or 'scout', the razor bill or 'marrot' and the puffin or 'willick'. In hard winters the fulmar or 'malmoch' appeared in the Forth but left in spring; the storm petrel often turned up in Leith harbour and the northern diver was to be seen in the Forth. The items on sale in the poultry market added to the variety. The most notable species was the 'solan goose' or gannet. Young birds were taken every year from the Bass Rock for which a considerable rent was paid. They were generally brought to market from the end of July to about mid-September when the whole colony left the Rock to spend the winter at sea, returning in early May. In winter the supply of different species of wild duck was generally quite large, but not as constant as in England since catching duck was not a business in Scotland. Mallard, wigeon, teal and golden-eye were the commonest species for sale, often accompanied by the long tailed duck or 'caloo' and less often, the velvet scoter, caught by the fishers at Newhaven where they were known as 'sea-jucks'. Shags occasionally appeared on the stalls. Wild geese were commonly sold, especially the grey lag, the white fronted, the bean, the brent or 'horra' goose, the barnacle and the 'claik' or 'clek' goose. After winter storms mergansers and dusky grebes were expected. Game in season was abundant and included black cock or 'heath fowl', red grouse or 'muir fowl', ptarmigan or 'white fowl', partridge, woodcock, curlews or 'whaaps', and snipes, as well as smaller birds such as fieldfare, redwing, sandpiper, blackbird etc. During the summer eider duck or 'dunter goose' or 'calk', and sheldrake or 'sly goose' appeared. Even the bittern or 'bog-blutter' or 'bog-bummer' was occasionally sold. Other common items included the coot, water-rail, golden plover, the lapwing or 'peasiweet' or 'teuchit' and especially pigeons, both common and wood pigeons. Rabbits were also sold in large number in the poultry market together with hares in season. The rabbits were caught in the warrens at Gullane in East Lothian.

Neill was a regular visitor to the large fish market since it was an excellent source of information about the inhabitants of the Forth. Located under the arches of the North Bridge adjacent to the Green Market, it was bordered by stalls chiefly occupied by salmon, trout, sea trout and char. The salmon were drawn from a number of different rivers while the trout and char were from Loch Leven and the sea trout from the mouth of the Esk at Musselburgh. Pike and perch were from the Lake of Linlithgow. The centre of the market was occupied by rows of Newhaven fisherwomen with their wicker baskets. They were well known for their habit of demanding a ridiculously high price for their wares but to the accompaniment of hard bargaining – conducted in the broadest dialect – they would eventually settle for about a third of the initial price.

Sea fish such as smelt and sperling were on sale in March and April when they came into the Forth in millions. Cod and haddock were almost always on sale in abundance. During winter great quantities were brought by cart from Dunbar and Eyemouth but the supplies from Newhaven and Fisherrow were preferred since the carriage did not improve their condition. Ling was less common than cod and was sold at a higher price. Whiting – often of large size – was regularly sold. Eels were on sale but not much in demand. Coal fish were common – sometimes as large as a full grown salmon – in which case they were known as 'sathes', 'says' or 'grey lords'. Pilchards appeared in quantity in October or November, followed by the herring which continued until March. In May and June shoals of sprats or 'garvey herring' appeared in the Forth and were on sale during the summer but seldom in large numbers. The sea-cat or wolf fish was uncommon but despised on account of its name by those unaware of its culinary excellence. Flat fish were abundant. The halibut (known locally as turbot) and the true turbot or 'rowan fleuk' was more or less common throughout the summer. Sole – rather infrequent and of small size – came from Aberlady Bay in July and August. Plaice, dab and flounder appeared throughout the year and were sold indiscriminately as 'fleuks', although small plaice were sometimes distinguished as 'salties'. Under the name 'skate' were included true skate and the thornback – the most esteemed species – and also the sharp-nosed ray. The young of all these species were known as 'maiden skate'. Other common but not so popular species included the father-lasher or 'lucky- proack' and the grey gurnard or 'crooner', so-called because of the purring sound caused by air forced through the gills. The blenny or 'greenbone' – as well as sand eels – appeared on the stalls in summer.

The gar fish was occasionally caught. The saury-pike or 'gowdanook' – hardly known in southern Britain – sometimes entered the Forth in immense shoals. The smooth hound, the tope and the angel fish were sometimes caught in the nets and taken to Newhaven to obtain oil from the livers by boiling. The piked dogfish accompanied the herring shoals and were a nuisance since they damaged the nets. Although of no economic importance, Neill was interested in the gaily hued dragonet, known locally as 'gowdie' or 'chanticleer'. One or two sturgeon might appear for sale during the season. Lobsters caught off the Fife coast were regular features, while nephrops caught at the mouth of the Forth were only occasional. Crabs were caught in vast quantities in spring and early summer. Often only the giant claws were brought to market. Oysters appeared on the stalls from September 1st to May 1st. Dredging for oysters provided a living for many Newhaven families. The close season for dredging was regulated by the Edinburgh magistrates. Clams or scallops and razor shellfish or 'spouts' were often on sale. Large quantities of common mussels were gathered by the fisherwomen at ebb tide and met with a ready sale. Another species from deep water was used chiefly for bait but sometimes brought to market. Cockles and limpets were neglected but whelks or 'buckies' were hawked through the streets, along with 'dulse' and 'tangles' – the blades of *Fucus palmatus* and the tender stalks of *Fucus digitatus* respectively. Another of the evening street cries was 'fine prawns' which were caught together with a few shrimps from the sandy bay at Portobello.

Before the removal of the tax on salt, itinerant women in the city streets mostly sold it from creels carried on their backs, although it was also available in shops. They came mostly from the salt pans at Fisherrow but some came from as far away as Prestonpans, nine miles away. They arrived every morning and returned home the same day.

Neill was understandably enthusiastic about the plants of the district. An excursion of a day or even a few hours could fill the botanist's collecting box with 'no contemptible spoil'. Even Arthur's Seat boasted about 400 different species, including some uncommon ones such as the spleenwort (*Asplenium septentrionale*), the sandwort (*Arenaria verna*), the potentilla (*P. verna*), the salvia (*S. verbenaca*), the meadow-rue (*Thalictrum minus*) etc. In the appendix to Lightfoot's *Flora Scotica* a Mr Yalden had listed the species to be found in the King's Park in Edinburgh, although he had not included such species as sanicle, the yellow mountain pansy and juniper – not to mention many mosses, algae and fungi. Other places to find interesting plants were the Pentland Hills, Habbie's How, Woodhouselee and Colinton, where there was a stand of the valerian *V. pyrenaica*, which Neill thought indigenous (although it is now known to have been introduced). Further afield Neill advised a search along the shores of the Forth for species like Scots lovage, sea rocket, prickly glasswort and purple cockle head, while offshore there were more than 30 species of seaweed, especially what were then known as species of *Fucus*. Thus in the space of his short article, Neill included a representative sample of species, sufficient to whet the appetite of any visiting botanist.

It is also worth mentioning that Neill also drew attention to the molluscs (both terrestrial and aquatic) and insects of the district. He also included a substantial section on the area's geology with a summary of the main features of the terrain and a concentrated account of the principal minerals and rocks. He recommended the reader to Professor Jameson's articles in *The Edinburgh Philosophical Journal*.

Much later in 1882, William Evans – writing of the mammals of the Edinburgh district – noted some of Neill's records from the first decade of the century.[4] Neill had found the water shrew in the River Esk in Habbie's Howe near Carlops. He recorded the polecat or 'foumart', squirrel and black rat among the fauna. This information was reported remarkably enough, in a two-volume work entitled *The Gentle Shepherd* by Allan Ramsay (1808).[5] This provided a detailed commentary on the poem, a glossary of Scots words and lists of animals and plants that Neill had observed at several sites which were mentioned in the poem. Some of the lists were quite impressive. At the site known as 'The Washing Green' Neill listed 26 species of mammal, 81 birds, five reptiles and amphibians, eight fish and just over 100 wild plants both native and introduced. Evans also noted that the eminent Swiss naturalist de Saussure visited Scotland in 1807 and was accompanied by Neill in visits about Edinburgh, where he saw great numbers of wild rabbits in the dunes behind the village of Gullane, by the Firth of Forth. Evans also referred to Neill's reports of whales in the Forth. His description of the stranded beaked whale has already been noted but he had also encountered killer whales or grampuses. In October 1814 a school of these animals appeared in the Forth estuary. Fifteen were killed at the mouth of the River Devon, two more at Tullibody and a further two near Stirling. Neill had noted that they were of both sexes, ranged from 9–20 feet in length, with shining black skin, white belly and eye patch and 24 beautiful teeth in each jaw.

Another marine curiosity – this time from the Shetland seas – was reported in the first volume of *The Edinburgh New Philosophical Journal*.[6] Several years previously in 1820, Neill had received from his friend Robert Strong of Leith, a specimen of a large and uncommon fish which had been sent to him along with the usual cargo of dried fish from Shetland. It had been split and cured in the usual manner. The head was fairly intact but other parts had undergone some decay. Attempts to preserve it proving unsuccessful, Neill had shown it to Professor Jameson, Dr Fleming and other naturalists and was only persuaded to publish an account because the species had not yet been admitted to the British fauna. It was just over five feet long from snout to tip of tail, with a disproportionately large head. He identified it as *Sciaena Aquila*. The fish was caught off Uyea in November 1819. It was first seen 'in contention with a seal' or rather, endeavouring to escape from its attacker. Some men sailed out to it and captured it without difficulty, since it was exhausted by the long struggle. It had suffered a bite about the mouth. The fins were of a beautiful red colour and the flesh white and very soft. No one had seen a fish like it before. When brought into the boat it emitted a buzzing sound. Neill had no doubt this was the *Sciaena* species described by Cuvier and known to the French as *maigre* or *aigle de Mer*. He mentioned that the buzzing sound or 'mugissement' of the *maigre* was louder than that of the gurnard. In present day accounts this species – referred to in English as 'meagre' or shade-fish – is known to have a very wide tropical distribution in the Atlantic and Indian Oceans. It occasionally turns up in British waters but not often as far north as Shetland. Neill noted that it had formerly appeared commonly on the French market but had disappeared for many years.

The siren

In 1828 Neill read a paper to The Wernerian Society about his experiences with a specimen of the tailed amphibian *Siren lacertian*. The description was later printed in the fourth volume of *The Edinburgh New Philosophical Journal*.[7] In Neill's day the taxonomic status of this animal was uncertain since it possesses external gills as well as lungs, and lacks hind feet. A sluggish inhabitant of muddy swamps in South Carolina and neighbouring States, little was known of its habits. Because of the external gills, a number of (especially Italian) zoologists – and also the Scottish zoologist, the Rev. John Fleming of Flisk – thought it to be at a larval stage, although Linnaeus and later Cuvier had concluded correctly that it was not. In 1825 Professor Monro (of anatomy and surgery in Edinburgh) received in a small barrel of mud and water a live specimen from America, measuring about a foot and a half in length. He handed it over to Neill for safe keeping. To provide a home for the latest addition to his little menagerie, Neill had a box constructed with an inclined plane that provided the animal with the option of resting either in water at the foot of the box, or in air on the sloping board. Food was provided in the form of small worms or sticklebacks. At first the box was kept in the greenhouse. A year or so later it was transferred to the hot-house at a temperature of 60–80°F. Here it became more active, sufficiently so to work its way out of the box and falling three and a half feet onto the greenhouse floor, thereby disproving the two current misconceptions that it would die if removed from water and break into fragments if dropped. It survived its adventure and continued to grow in size. Just how long it survived with Neill is unclear but it certainly lived at least six and a half years.

Seaweeds

Neill wrote an article on seaweeds for Brewster's *Edinburgh Encyclopaedia*.[8] He introduced the topic with a discussion of alternative approaches to classification by several authors, together with general comments about structure, colour, distribution and the practical uses of seaweed for food and other purposes. The different species were then classed within the genus *Fucus*, although several were later renamed as indicated below.

In the north of Scotland *Fucus loreus* (*Himanthalia elongata*) was used to make a kind of ketchup for use with fish or fowl. *F. esculentus* (*Alania esculenta*) – an edible species – was known on the east coast of Scotland as 'badderlocks' and in Orkney as 'hen-ware'. The esteemed dulse of lowland Scotland *F. palmatus* (*Palmaria palmata*) and tangle *F. digitatus* (*Laminaria digitatus*) were both eaten fresh from the sea, without preparation. As noted above 'dulse and tangle' was one of the Edinburgh street cries. Although sometimes regarded as a salad they were more often classed as a whet. Dulse was no longer fried. *F. saccharinus* (*Saccharina latissima*) from the Forth made but a wretched pot-herb. *F. digitatus* (*Laminaria digitatus*) was sometimes used to make handles for pruning or grafting knives. The thick stem was cut into four-inch pieces into which were inserted the handle of the knives. As the seaweed dried it became hard and closely invested the knife handle. In the north of Scotland – especially Orkney and Shetland – the large stalk of this seaweed were used for fuel, while *F. vesiculosus* was used as cattle feed. The kelp of commerce was derived from eight or nine different species, but especially *F. vesiculosus*, *Lamiaria digitatus* and *F. nodosus* (*Ascophyllum nodosum*).

Hail in Orkney.

In the fourth volume of *The Edinburgh Philosophical Journal* Neill gave a graphic account of a remarkable hailstorm that had occurred at Stronsay in Orkney on July 24 1818, as reported to him by the local minister the Rev. Mr Taylor.[9] There was an immense fall of hail, accompanied by thunder and sheet lightning. The fragments of ice were of many shapes and sizes. Some of the hailstones were as big as goose eggs and fell with such force as to become embedded several inches into the soil. But there was also a great quantity of angular pieces of ice – some as big as oyster shells – which covered the ground to about nine inches, piling up to a foot and a half in the farm courtyard. All the farm windows facing the storm were shattered and the crops were flattened and quite destroyed. Cattle in the path of the storm were terrified – many were cut and bleeding and some would not survive. Tame geese feeding on the links were in a pitiable state. Only a few were still living after the storm passed. Bills were split and many had their eyes knocked out of their sockets, hanging by the nerve. The layer of shattered ice was so thick – walking across it was practically impossible for fear of slipping and being cut about the legs by the sharp pieces of ice. This most unusual storm was very localised and soon passed but left havoc in its wake.

1   Lyell 5. Letter from Neill to Charles Lyell, the Geologist, about the 'Sea Monster', March 14 1848, Edinburgh University Library.

2   Patrick Neill, *Account of a Forfar Garden*, 1809, *The Scots Magazine*, Vol. 71, pp. 409–410.

3   J. Stark, *Pictures of Edinburgh Containing a Description of the City and its Environs*, 6th edition, 1834, John Anderson Jun., A. & C. Black and W. Hunter, Edinburgh.

4   William Evans, *The Mammalian Fauna of the Edinburgh District*, 1892, McFarlane & Erskine, Edinburgh.

5   Allan Ramsay, *The Gentle Shepherd. A Pastoral Comedy with Illustrations of the Scenery*, Vol. 1, for William Martin, W. Creech, A. Constable & Co., F. Hill & A. Mackay, Edinburgh, and John Murray, London. This volume contains lists of animals and plants from different places by Patrick Neill.

6   Patrick Neill, *Account of a Rare Fish (Sciaena Aquila) Found in the Shetland Seas. The Edinburgh New Philosophical Journal*, n.d., Vol. 1, pp. 135–141, read before the Wernerian Natural History Society, May 27 1826.

7   ----------, *Some Account of the Habits of a Specimen of Siren Lacustris Which Has Been Kept at Canonmills Near Edinburgh For More Than Two Years*. Read before The Wernerian Natural History Society, Jan 12 1828. *The Edinburgh New Philosophical Journal*, Vol. IV, pp. 346–355.

8   ----------, *Fuci, The Edinburgh Encyclopaedia. Conducted by David Brewster with the Assistance of Gentlemen Eminent in Science and Literature*, MDCCCXXX, Vol. X, pp 1–23.

9   ----------, *Account of a Remarkable Shower of Hail Which Fell in Orkney on the 24th of July, 1818, The Edinburgh Philosophical Journal*, Vol. IV, from October 1 1820 to April 1 1821, pp.365–371.

5 ❧ The Wernerian and Caledonian Societies

The years 1808 and 1809 witnessed two events which determined how Neill would spend much of the rest of his life. In 1808 he became a founder member and secretary of the Wernerian Natural History Society and the following year, a founder member and secretary of the Caledonian Horticultural Society. He continued in office as secretary to both societies until shortly before his death in 1851. His key role in these rather different organisations won him friends and acquaintances among the Scottish scientific intelligentsia, as well as the landed gentry and their head gardeners who were the mainstay of the horticultural society. That he served so long is sufficient testimony to the conscientious way he looked after the affairs of both societies.

The Wernerian Society owed its inception to Robert Jameson, who held the Chair of Natural History at the University of Edinburgh for half a century. At a meeting on January 20 1808 of the persons who had been encouraged by Jameson to form a scientific society, a constitution was agreed and office bearers appointed.[1] These comprised the president – who of course was Jameson – four vice-presidents, eight councillors, a secretary – namely Patrick Neill – a treasurer, librarian and Patrick Syme, the society artist. The three honorary members comprised Abraham Gottlob Werner (a German geologist), Sir Joseph Banks (president of the Royal Society of London) and Richard Kirwan (president of the Royal Irish Academy). Members of the society – admitted by election – were distinguished as resident, non-resident and in later years as corresponding members. Foreign members included Baron Cuvier of Paris, Baron von Humboldt of Berlin and Louis Dufresne. Many of the leading lights of the Scottish scientific community were members such as David Brewster, Robert Knox, Sir William Jardine, and Dr T. S. Traill etc. During his stay in Edinburgh the American bird painter Audubon – with Neill's active support – was elected a member to his great delight. Meetings were held monthly except during the summer, when they were occasional. Guests could be invited and non-members could submit a paper to be read at a session, but only by an existing member. The new society quickly established a reputation which spread as far as America. To be elected a member of the Wernerian Natural History Society was a prized recognition in the scientific community of Edinburgh and beyond. At the height of its influence there were some five hundred members, although the majority of them were non-resident. The scientific papers which were read to the society might be published in full or in summary form. A number of the more substantial papers were published in eight occasional *Memoirs of the Society* or in other journals such as *The Edinburgh New Philosophical Journal* – of which Jameson was the editor – or *The Scots Magazine*. The proceedings were often reported in the local press.

Although the term 'natural history' appeared in the title of the society and many of the papers which were read dealt with some aspect of plant or animal life, others were devoted to geology (especially in the early years), mineralogy, geography, meteorology, physics, exploration etc. For example Volume I of the *Memoirs* included a paper by Jameson on the mineral system of Werner, an account of animals collected during the arctic expedition of Sir John Richardson, a history of fishes found in the Forth and nearby rivers and lakes by Neill, and an account of a stranded fin-whale, also by Neill. Volume II included an account of sponges found around the coast of Britain, an article on the Greenland polar ice by W. Scoresby Jnr., and a paper in Latin by W.E. Leech on the arrangement of oestridous insects. Volume III included observations on the orang-utan by Dr T.S. Traill, a paper by Dr Brewster on the axes of double refraction in crystals and a paper on the bed of the German ocean by Robert Stevenson (Robert Louis Stevenson's grandfather). Volume IV included an account of new species of Scottish fungi by R. K. Greville, some observations on the natural history of the mole by the Rev. J. Grierson, and a paper on the ability of spiders to launch themselves into the air by means of their silken threads by John Murray. In Volume VI Audubon – before he was elected a member – submitted a paper on the habits of the turkey, read to the meeting by the secretary. Later – after election – Audobon dealt with the habits of the alligator and the rattlesnake. In the same volume in 1828, the Rev. Dr Scott of Corstorphine read a paper which illustrated the extent to which biblical accounts were accepted at their face value at that time. He argued that the great fish that swallowed Jonah and cast him up three days and nights later on dry land could not have been a whale as commonly supposed, but must have been a whale shark. In more scientific vein Dr Robert Knox (the anatomist) described the development of the poison fangs of the serpent. Evidently there was no obvious limit on what could be brought for discussion before the society. The last volume contained only two very long papers on the economic value of fishes in the Forth district and the geology of the Lothians respectively. By then the society was beginning to run out of steam. Membership was declining. By the 1850s age and ill-health were overtaking Jameson who died in 1854. In 1858 the society was wound up and that was the end of what had once been a prestigious and scholarly association, now so forgotten that its name does not even appear in the index of a recent encyclopaedia devoted to Scottish history and affairs.[2]

There are two aspects of the Wernerian Society which call for comment. Since the subject matter of the meetings overlapped those of the Royal Society of Edinburgh, and the majority of the office bearers – including Jameson and Neill – and many of the other members were already fellows of that society, one might question the need for another rather similar society in Edinburgh. The clue lies in its name. Why should an Edinburgh scientific society be named after a German geologist? There can be no doubt that the name was chosen by Jameson, who in 1800 had been a student of Werner at the Bergakademie in Saxony, where he attended lectures, visited mines and field sites until 1802 when he returned to Scotland. At that time there was a great divide of scientific opinion about the origin of the rocks of the earth's crust between those who – like Werner – attributed a primary role to the action of water and the precipitation of sediments and those who considered heat and volcanic action as key agents. The two schools of thought – dubbed respectively Neptunists and Plutonists – displayed the kind of rivalry and fervour commonly associated with football teams. Werner was a leading Neptunist in the European scene, in opposition to the interpretation of rock formation described in *Theory of the Earth* in 1788 by the Edinburgh geologist James Hutton, whose views were endorsed and extended in John Playfair's *Huttonian Theory of the Earth* of 1802. Perhaps Jameson – as a disciple of Werner – thought the Royal Society of Edinburgh suffered from too much Plutonist influence. After all Sir John Hall – a prominent Plutonist – was president. By forming a new society named after Werner, Jameson was stealing a march on his rivals. In the long run Hutton and the vulcanists carried the day and even Jameson had to accept it.

It is important to note that right from its inception, there were those who considered the name 'Wenerian' inappropriate. In a letter to J.E. Smith, the President of the Linnean Society in 1808, Neill mentioned that the old Edinburgh Natural History Society was still active[3]. Smith had been a founder member when a student of medicine at Edinburgh. Neill was very familiar with its affairs since he had been a secretary for 'two years back'. He claimed to have nearly effected an amalgamation of the two societies. The name 'Wenerian' had given 'some offense'. However, the Edinburgh Natural History Society had for many years been a debating society for students and gentlemen who had not finished their education. The membership of the two societies was too different to be amalgamated and they could hardly change the name of the new society without incurring ridicule.

Neill occasionally contributed papers to the society. Interested in all things marine, in 1808 he wrote about the fish to be found in what he always termed 'the Frith of the Forth', and nearby rivers and lakes.[4] He did not claim it was a complete account but covered the species he had encountered in recent years. This was a substantial, well-referenced article. He was a regular visitor to the fish market and kept in touch with the Newhaven fishermen. When he enquired about specimens of a small fish known as the dragonet that he wished to dissect, he was inundated with them. As was his wont, he noted the Scottish names for wild species. Thus the dragonet was also known to the fishermen as 'chanticleer' and 'goudie'. The basking shark – called 'sail-fin' or 'cairban' in the Western Isles – was known as 'pricker' or 'brigdie' in the north of Scotland. In a paper by Jameson on the class Vermes found in the Forth and elsewhere in Scotland, Jameson acknowledged receipt of specimens of marine worms he had often received from Neill, who kept an eye on what was brought to the surface by the oyster dredgers.

In another paper Neill reported his observations on a fin-whale stranded on the banks of the Forth near Alloa on October 23 1808.[5] Farm servants nearby had been awakened by the noise of blowing and floundering of the whale. It was November 1st before Neill could get to the carcass. It had already been stripped and cast afloat on account of the stench only to come ashore again about two miles down on the opposite bank. Mr Robert Bald – civil engineer and manager of the local colliery and long-term friend of Neill – had measured the dimensions and reported the length as 43 feet and the circumference at the thickest point as 20 feet. The blubber only about two inches thick, filled seven large casks, but was of poor quality. The total was sold to a soap boiler for £15. Neill could not help observing that if it had been combined with peat moss, as advised in a recent paper *The Transactions of the Highland Society* by Lord Meadowbank, it would have been of greater value as manure.

In Volume II of the memoirs there was a joint paper by Dr Barclay and Neill about the fate of a Beluga whale in the Forth.[6] In June 1815 Neill received a letter from Mr Bald who reported that, for about three months an animal of singular appearance and white in colour had been seen passing and repassing the harbour at Alloa. Neill set off at once and confirmed the description. Many attempts had been made to kill the beast by attacking it with guns and spears. It was finally killed by salmon fishers off the Abbey of Cambuskenneth. The animal proved to be a Beluga whale of 12 feet four inches in length. Mr Bald purchased the carcase and sent it to Professor Jameson. Dr Barclay dissected it while Mr Syme – the society artist – made a painting of it for display at the next meeting. The contrast between the public attitude to whales at that time and that of today could hardly be greater.

In a later paper Neill gave a valuable account of the fossil remains of the beaver found in Perthshire and Berwickshire, proving that the animal was formerly native to Scotland.[7] He referred to 9th Century Welsh records of the prices of skins of marten, otter and beaver, in which the relatively high price of beaver skin showed the animal was already scarce. There was further evidence of its existence in Wales from the 12th Century. In Scotland Hector Boece writing at the end of the 15th Century, referred to the beaver in a matter of fact way. Neill had found no mention of the beaver in subsequent public records. Robert Sibbald – author of *Scotia Illustrata* (1654) – had merely quoted Boece and failed to comment or 'show any of that precision and zeal for enquiry which characterises many other parts of his writing'. Neill had contacted Dr Stuart of Luss – both naturalist and Celtic scholar – who confirmed that the ancient gaelic name for the beaver was still recalled in remote, western districts of Scotland. It was known as 'losleathan' (from 'los' – a tail) and 'leathan' (which signified 'broad', very similar to the Welsh *llosdlydan*). Dr Stuart was aware of a tradition that the beaver once occurred in Lochaber. Neill then dealt with the fossil evidence. Many years previously he had learned from the minutes of the Society of Antiquaries of 1788 that a Dr Farquharson had presented to the society the fossil of the head and the haunch bones of a beaver, dug up in Perthshire and still preserved in the society's museum. He was able to examine it and further reported that it was found under five to six feet of peat moss in one of the marl pits of the partly drained Loch of Marklee, in the parish of Kinloch near the foot of the Grampians. A pair of large deer horns was found nearby.

In 1818 in the parish of Edrom, Berwickshire, the proprietor of the estate of Kimmerghame drained a morass called Middlestot's Bog. Under the peat moss in a layer of shell-marl, the remains of a beaver were found at a depth of seven feet. The bones of the skeleton had fallen apart due to uneven subsidence. Neill identified some of the mosses in the peat and with a naturalist's acumen, noted that the presence of shells of the fresh water molluscs *Limnaea* and *Succinea* suggested the presence of a long period of still water. The skull and lower jaw were presented to Professor Jameson for the college museum. Here again a pair of large deer horns was found nearby.

Neill then compared the fossil bones with bones of beavers from Hudson's Bay to confirm they belonged to the same species – although the two Scottish specimens were more alike to each other than to the Canadian specimens. Finally he noted that the Scottish specimens were closely similar to the illustrations of fossil beaver described by Cuvier from a specimen found in a peat moss in the valley of the Somme, where big deer horns were also found.

A Sour Note

We cannot leave the Wernerian Society without mentioning an extraordinary attack on Jameson's presidency. In 1830 the serene ambience of Edinburgh's scientific community was ruffled by an uncompromising criticism of the conduct of Professor Jameson as President of the Wernerian Society, by former student Henry Hulme Cheek, in the pages of the recently established *Edinburgh Journal of Natural and Geographical Science* (of which Cheek was one of the editors).[8][9] Although unsigned, there was no doubt about who wrote it. This set in motion an ongoing dispute with Neill who, as secretary of the Wernerian Society, rushed to defend his friend Jameson. There were angry exchanges of letters, a further attack by Cheek in the pages of his journal, a face to face row between Cheek and Neill, accusations of false reporting of facts, an intervention by Dr Knox the anatomist and, finally, a weighty pamphlet by Neill which brought the altercation to a close.

The general pros and cons of the incident can be stated briefly as follows. Cheek had a number of criticisms. In spite of having existed for 20 years, why were the *Memoirs* of the society so few in number? Cheek had an explanation. Jameson had been president of the society since its foundation. He had also founded and edited *The Edinburgh New Philosophical Journal*, which often published papers that had been read to the Wernerian Society. Cheek took this as evidence that Jameson was siphoning off the best Wernerian papers to bolster the fortunes of his *Philosophical Journal*. He also strongly objected to the way the society's library was run, apparently to serve Jameson's interests and convenience. During his few years residence in Edinburgh Cheek had been grieved to see the University Museum closed to those students who did not purchase at exorbitant cost, 'the notional privilege' of access. He objected that men of science could only visit the museum with Jameson's permission. Why had the Wernerian Society no rules – like other societies – which set limits to the period of service of office bearers? Although he had only respect for the members of the Wernerian Society, it was in the hands of a coterie that brooded over it 'like a night mare'. At first Cheek exonerated Neill – whom he conceded was universally held in esteem – but when Neill counterattacked, he ridiculed him for his defence of Jameson and concluded he must be included in what was wrong with the society. Cheek ended one of his later diatribes with the assertion he would if necessary willingly do again what he had done, provided he could 'maintain that spirit of disgust by which I am now led to point my finger at a sickly and overgrown monopoly'.

Neill took up the cudgels in an equally determined manner[10] [11] In his pamphlets he utterly rejected the accusations of dishonesty levelled at Jameson, some of which were actionable. The rules of admission to the library had been freely agreed by the Wernerian Society. The criticism over the paucity of the *Memoirs* arose from ignorance. The society was short of funds and could not afford more frequent numbers. It had been agreed that papers read to the society could be published elsewhere, if the authors agreed. Indeed they were glad to see their work published in *The Edinburgh New Philosophical Journal*. Neill was astonished at Cheek's attitude. When Neill first knew him he had taken him for a diligent student of nature, had told him he could apply for tickets to the Wernerian meetings and had given him letters of introduction to fellows of the Linnaean Society in London. Cheek had denied any particular acts of kindness towards him and complained that he had never been invited to Neill's home. So – said Neill – he evidently wants kindness in the shape of 'beef and pudding'. Cheek bridled at being referred to as 'a young stranger'. Since when had an Englishman been regarded as a stranger in Edinburgh? But he had himself claimed the role of 'Censor of Edinburgh', since as an outsider 'he was not shackled by local connections'. Neill acknowledged the indignity of becoming embroiled in this paper-war but he was anxious to protect the membership of the society from getting involved in the controversy, so he had decided to set the record straight on his own. One suspects that he rather relished a verbal scrap since he was no mean exponent of sarcastic invective.

When the protagonists at length fell silent what are we to make of this row? We can feel some sympathy for Cheek who was certainly right in his criticism that the Wernerian Society lacked rules to limit the terms of service of the principal office–bearers but he pursued his vendetta against Jameson with all the unqualified abandon of youth. Jameson evoked different reactions in different people. To some he was warm and forthcoming, to others cold and authoritarian. It is well known that Charles Darwin, when a student at Edinburgh, found Jameson's lectures so boring they put him off geology for years. Certainly Jameson took himself seriously – witness the title page of *The Wernerian Memoirs* listing his membership of learned societies all over the world in 20 lines of small type. But then he worked tirelessly and successfully to establish the University Museum; he was always short of funds so the admission charges to the library were welcome; he was a good field geologist and an acknowledged expert in the field of mineralogy. He displayed complete indifference to the attacks upon him. When the dust settled nothing had changed. He and Neill remained in charge of the Wernerian Society. Cheek's journal soon ceased publication while Cheek himself disappeared from the scene.

The Caledonian Horticultural Society

Unlike the Wernerian Society, the Caledonian Horticultural Society survived to become a permanent feature of the Edinburgh scene and is still both active and influential. Fletcher and Brown have described how on November 25 1809 Thomas Dickson (the leading Edinburgh nurseryman) called a few like-minded friends together to consider the formation of a Scottish horticultural society.[12] There can be no doubt that this initiative was prompted by the formation of the London Horticultural Society five years previously. The proposal was attractive, so later the same year professional and amateur gardeners were invited to a meeting in the Physicians Hall, George Street, to ascertain the degree of interest in the formation of such a society. The general reaction was again favourable to the idea and led to its formal establishment. This crucial meeting was chaired by Dr William Duncan (the Senior Professor of Medicine) who was a kind of father figure to the society in its early years. A council was voted in with the Duke of Dalkeith as president and Patrick Neill and Walter Nicol as joint secretaries, to launch what proved to be a successful venture. Walter Nicol had been gardener to the Marquis of Townsend in Norfolk and then gardener at Wemys Castle before taking up the writing of popular books on practical gardening. Sadly he died in 1811 leaving Neill to carry on alone for almost the rest of his life, a total of 40 years as secretary.

The council of the society was well endowed with academics that included not only William Duncan but also the professors of *Materia medica*, botany and agriculture. Other members included professional nurserymen (such as two of the Dicksons and Edward Sang of Kirkcaldy), head gardeners (like James Macdonald of Dalkeith), landowners (such as Sir George Stuart Mackenzie of Coul, Sir Robert Liston of Millburn Tower and Sir James Hall of Dunglass), as well as private gardeners – so the foundations were soundly representative.

The new society proceeded forthwith to a programme of activities. It held regular meetings at which members read papers on all manner of horticultural topics, especially on the management of fruit trees. Head gardeners from estates in southern Scotland were the chief contributors.[13] The best papers were published in a series of *Memoirs* which were continued until 1829 when a different form of reporting was adopted. The society also quickly established competitions – especially for orchard fruit – in which gardeners submitted baskets of named varieties of apples, pears, plums etc. There was brisk competition for first place and the modest prize which was awarded. There was great emphasis on the correct naming of the varieties. The extensive minutes – written up by Neill – record both the names of the varieties and of the gardeners who sent them in. Hence they provide a valuable record of the most popular varieties of orchard fruit and how they changed with time. Between 1809 and 1858 altogether 156 names of apples and 90 of pears were recorded. In later years ornamentals entered the lists and orchard fruit dwindled in significance.

It was hoped that the new society would be favoured with a Royal Charter. When Neill applied for it his application was rejected because it was thought the membership was too localised and parochial. However, Neill pointed out that a substantial number of members were resident in different parts of the British Empire, so the society had far-flung support. This evidence won the day and in 1824 the Caledonian Horticultural Society added 'Royal' to its title.

An important development – dear to Neill's heart for he had advocated the need for it before the society was formed – was the establishment of an experimental garden. In 1817 at the earlier suggestion of Sir John Sinclair, the society sent a delegation consisting of Neill (who led it), John Macdonald of Dalkeith and the garden designer John Hay to survey the state of horticulture in the Low Countries and Northern France. Their three-month tour is described later. In addition Neill made a trip to France with a similar intent in 1821. The principal reason for these excursions was to become familiar with the latest horticultural methods in anticipation of the formation of the Experimental Garden – the foundations of which were laid in 1822 when the society entered into an agreement with the Lords of the Treasury and obtained ten acres of ground adjacent to the Royal Botanic Garden.

In November 1824 Neill sent out a circular to members explaining the need to raise funds for the project, since the existing income was insufficient to establish the Experimental Garden and maintain it. At a general meeting of the society the following was agreed: to offer shares at 20 guineas each; to charge those ordinary members who were not shareholders one guinea yearly for the benefit of the Garden; to charge an admission fee of two guineas and appeal for voluntary subscription.

In March 1825 the minutes included a report on the progress of the Experimental Garden, consisting of an ambitious plan which very likely owed not a little to what Neill hoped for. There was to be an orchard of two and a half English acres, sufficient to accommodate 550 trees of apples, pears, plums and cherries. There were to be separate plots dedicated to the raising of different kinds of stocks for budding or grafting fruit trees, a brick faced wall 14 foot high by 450 feet long for fine French and Flemish pears, apricots and peaches, a walled garden for the naturalisation of tender exotics, an enclosed plot devoted to experiments, and other plots for ornamental plants or strawberries as well as one for agricultural grasses and clovers together with a pond surrounded by rock work for alpines etc. There was to be room for bush fruit as well as a hot-house and forcing houses for melons and pineapples. In short, it was to be an exercise in the most up-to-date arts of horticulture.

Although it appears that due to shortage of funds, the full programme outlined in Neill's minutes was not realised yet a great deal was accomplished, especially after the appointment of James McNab (son of William McNab – former superintendant of the Royal Botanic Garden Edinburgh), as superintendant in 1836. His drive led to the erection of an exhibition hall funded by voluntary subscription, a Winter Garden hall, a camellia house and an impressive collection of plants from all over the world. From 1842 we have his list of 142 different sorts of apple and some 90 sorts of pear almost entirely of French or Flemish origin, growing in the Experimental Garden.[14]

The minutes reveal that Neill periodically had to chase up members who were in arrears. He drew particular attention to the benefits they enjoyed from the existence of the Experimental Garden. The finest new varieties of all the fruits had been obtained from reliable sources and extensively propagated to supply the wants of members with specimens, according to their contribution and order of applying. But the Experimental Garden fulfilled another function for it had become a fashionable rendezvous and one of the chief amenities of the city. The Garden had been originally designed to offer pleasing views and enjoyable walks, serving in that respect much the same scope for relaxation as the Royal Botanic Garden does at present. The Winter Garden was a real innovation for Edinburgh since it constituted a fashionable promenade, likened to the contemporary *jardin d'hiver* in Paris.

However, in due course the enthusiasm for maintaining the Experimental Garden waned. Membership declined and by the late 1850s the society was in financial difficulty. It was realised in the 1860s that in any case, the original aims and objectives had been achieved, since interest and appreciation of horticulture and gardening was now widespread and enterprising nursery firms abounded. It was therefore decided to make the Garden over to Her Majesty's Department of Woods and Forest, in return for a reasonable financial settlement. In 1864 the grounds were incorporated in the adjacent Royal Botanic Garden.

The Celebration of Neill's Service to Horticulture

In June 1843 when Neill had reached the age of 67, he was rewarded with a remarkable demonstration of the respect he had won among the gardening fraternity. A committee of 35 head gardeners from the major estates in the Lothians and Fife had been formed to organise subscriptions for a silver ornament and a testimonial for presentation to Neill. Subscriptions were sought among gardeners throughout Scotland, within the range of two shillings and sixpence and ten shillings. There was a tremendous response with nearly 600 gardeners subscribing. This was an exclusively Scottish affair since offers from English gardeners were declined. A testimonial bearing their names and the silver ornament were presented at a public dinner in the Café Royal Edinburgh, on June 22 1843. It was attended by a great many gardeners, to mark their high esteem for his personal character and the gardeners' appreciation of 'that zealous and continued devotion of his time and talents to the cause of horticulture'. The ceremony was described at length in the local press. After the loyal toasts, the chairman gave a glowing tribute to Neill, punctuated by frequent cheers of approbation. He observed: 'Many of the young men I see about me can bear testimony to his kindness, attention and courtesy. He is known worldwide, including America.' He ended his peroration on a pious note: 'When you at last close your eyes may you go to the great Parent of all and there enjoy the rewards of your works.'

Neill's reply was 'evidently under considerable emotion'. Ever modest, he thought his merits had been overrated. He had already received a number of rewards. In 1817 the Caledonian Horticultural Society awarded him a gold medal which he was wearing. In 1821 while he was abroad, the same society awarded him a piece of plate bearing an inscription by Sir George Mackenzie. The previous autumn the society had proposed that a marble bust by the most accomplished sculptor in the city should be placed in the new hall. And now to crown it all, there was this testimonial and splendid vase. He then paid a warm tribute to Scottish gardeners and horticulture. It used to be remarked by his friend Dr William Duncan – the father founder of their society – that gardeners and doctors were the most eminent exportations from Scotland. In general, four out of six head gardeners in England were Scots. Neill then moved a toast to James McNab – his right arm.

The piece of plate – now preserved in the National Museum of Scotland – was later bequeathed to James McNab, who presented it to the museum. It is a handsome silver vase standing two foot high on a triangular pedestal. The lid of the vase is surmounted by a figure of Britannia in the Scottish form. One side of the vase bears an excellent likeness of Neill surrounded by a wreath of flowers. The flowers on the wreath comprised three South

American species that were first figured in the botanical press from specimens grown in Neill's garden – and from there they were introduced to Britain. On the other side of the vase there are a further four species, also first figured from his garden. On one side of the triangular pedestal there are three female figures representing Flora, Pomona and Ceres. Below the medallion likeness of Neill there is the Neill crest, surrounded by a wreath of Scottish thistles, together with the likeness of the several plant species named after him. The vase was considered the most elaborate silver ornament ever produced in Edinburgh.

The remarkable testimonial from so many working gardeners and the public ceremony comprised fitting recognition of Patrick Neill's enduring support for all forms of horticulture and its practitioners. To recall the unprecedented event is to remind us of the progress in gardening and horticultural expertise achieved in Edinburgh during the first half of the 19th Century and Neill's benign influence in promoting it.

1   Minutes of the Wernerian Natural History Society, Library of the University of Edinburgh, Dept of Special Collections.

2   John Keay and Julia Keay (eds), *Collin's Encyclopaedia of Scotland*, 2000, Harper Collins.

3   Neill Letters, Linnean Society, London.

4   Patrick Neill, *List of Fishes found in the Frith of Forth and Rivers and Lakes near Edinburgh with remarks*, Memoirs of The Wernerian Natural History Society (I), 1809–1910, pp.526–555.

5   ----------, *Some Account of a Fin-whale stranded near Alloa*, Memoirs of the Wernerian Natural History Society, (1), 1809–1810, pp.201–214.

6   Dr Barclay and Mr Neill, *Account of a Beluga or White Whale killed in the Firth of Forth*, Memoirs of the Wernerian Natural History Society, (III), 1817–1820, pp.371–394.

7   Patrick Neill, *Account of some Fossil Remains of the Beaver (Castor fiber) found in Perthshire and Berwickshire, Proving that the Animal was Formerly a Native of Scotland*, Memoirs of the Wernerian Natural History Society 1819, (III), 1817-1820, pp.207–219.

8   Henry Cheek, *Comments on Wernerian Natural History Society*, Edinburgh Journal of Natural and Geographical Science, 1830, (I), pp.352–353.

9   ----------, *Comments on Wernerian Natural History Society*, Edinburgh Journal of Natural and Geographical Science, 1830, (II), pp.269–274.

10 Patrick Neill, *Address to Members of the Wernerian Natural History Society*, 11891–2, The Library of The Royal Botanic Garden, Edinburgh.

11 ----------, *Supplement to an Address to the Members of the Wernerian Natural History Society*, 11891–2, The Library of The Royal Botanic Garden, Edinburgh.

12 Harold R. Fletcher and William H. Brown, *The Royal Botanic Garden Edinburgh 1670–1970*, 1970, p.138, H. M. S .O.

13 Forbes W. Robertson, *Orchards, Fruits and Gardeners in Early Nineteenth-century Scotland*, Review of Scottish Culture 2008, No. 20, pp.43–56.

14 James McNab, *Scrapbook*, The Library of The Royal Botanic Garden, Edinburgh.

6 🌿 *Scottish Gardens and Orchards*

In 1813 Neill produced an extensive report on Scottish Gardens and Orchards, drawn up as he quaintly put it, 'By Desire of the Board of Agriculture'.[1] No doubt Neill was selected for this task by Sir John Sinclair, who played a large part in establishing the board in 1793. It was not until 1866 that the Board of Agriculture published figures of the acreage devoted to different crops. Although detailed statistics of the acreage of market gardens and orchards first appeared in 1875, only by 1877 were the published data for parishes and counties both comprehensive and reliable. The earlier evidence in the *First Statistical Account of Scotland* (1791–99), although of landmark status, varied in the scope and accuracy of the parish data. Hence Neill's historically important report comprised the first general survey of Scottish horticulture. Since he had a critical, scientific approach to evidence, the picture he drew of the scale and distribution of market gardens and orchards – as well as of the quality of domestic gardens of different classes of occupier in early 19th Century Scotland – can be accepted with confidence. On the title page he quoted his qualifications as a fellow of the Linnaean Society and one of the secretaries of the Caledonian Horticultural Society. It was a substantial report, running to 220 pages with numerous appendices. Although he used the term 'gardens', only edible produce was considered and apart from the odd comment, ornamental plants were not on the agenda.

In his introduction he briefly acknowledged the European origins of Scottish horticulture, mediated both by the monastic clergy – who quickly spread throughout the land after the Norman Conquest – and the influence of noblemen, who often travelled abroad and returned with ideas of how to ameliorate the often austere precincts of their ancestral homes. He attributed the general excellence of much of contemporary Scottish horticulture to the education widely provided by parish schools in Scotland, but less so elsewhere. He considered first the private gardens of different classes of society before dealing with market or 'sale' gardens, as they were then known.

As might be expected, the gardens of noblemen and gentleman with landed properties were the best managed. They excelled in the production of both the quality and variety of vegetables. Their kitchen gardens averaged three to five English acres, were usually larger than their English counterparts and might extend to eight or even 13 acres – as in the Duke of Buccleuch's garden at Dalkeith. Such kitchen gardens were so well kept that he had no criticism to make, unlike that levelled at many of the gardens of lesser mortals.

He provided a break-down of the annual cost of a typical gentleman's garden of five acres as follows:

	£	s	d
Head gardener's wages	60	0	0
Six and a half bolls of meal at 25s per boll	8	2	6
Cow keep during season	10	0	0
Coals for firing	3	0	0
House rent	4	0	0
Six garden lads at 11s. per week	171	12	0
Coals for firing	4	0	0
Lodging with bed	7	16	0
Two weeders, three months at 5s. per week	6	0	0
Kitchen ground requiring 20 carts dung per acre at 7s.	42	0	0
Tan for pine-stoves and pits, 40 carts	5	0	0
Coals for firing to the houses	30	0	0
Garden seed, glass, mats, tools, paint and nails	30	0	0
Cartage	35	0	0
Total	416	10	6

He wrote: 'By a careful calculation it was found that a grape-house 45 feet in length and 14 feet wide cost about £20 p.a., a narrower house of the same length £12 and a pine-stove, something about £30 p.a. including every expense.'

The gardens of feuars and owners of small properties were very often in good shape. The occupiers often saw themselves as 'little lairds' and sought to emulate the gentry in their gardens. As a rule, they grew all the basic vegetables such as cabbages, peas, beans, turnips, onions, leeks and carrots. As soon as the early crops were harvested their place was taken by winter vegetables such as savoys, greens and winter turnip. Sometimes there were a few cauliflowers and a little salad. There were often several rows of gooseberries, with red and white currants on the walls, together with a few apple and one or two pear trees.

Farmers' gardens, with few exceptions, were often of indifferent quality. Neill advised that the size of such gardens should correspond to the size of the family rather than the size of the farm, since it was better to manage a small garden well rather than a large one poorly. In the better gardens there was often a 12 foot wide border along the fence, perhaps 18 feet wide at the north end. The border below the wall was often planted with Virginian strawberries or curled parsley. There might also be a few paths for access to gooseberry, currant or raspberry bushes. The three-to-four foot walls were too low to keep out hares and rabbits. He favoured a three foot wall backed by a hawthorn and privet hedge for an effective barrier. Failing that, a seven foot wall would suffice. Some farmers kept a gardener but this was unnecessary unless the taste was for flowers or wall fruits, grapes or melons. Any of the young farm servants could soon learn how to manage a kitchen garden or a man kept for hedging and ditching could do the job.

The gardens rented by the year – belonging to day labourers, smiths, carpenters, tailors etc – contained a similar range of vegetables as found in farmers' gardens, but on a smaller scale. No article was more esteemed than the type of green of the cabbage family, known as 'stoans' or 'stowans'. In southern Scotland, German greens and savoys were planted early in the spring, but further north and in much of Perthshire, red curlies and plain red kale were preferred. The size of these gardens varied between three and 18 falls (96 and 574 sq. m.) with most between six and eight (191 and 255 sq. m.). The short period of the lease could act as a deterrent to improvement, since a better maintained garden could induce a potential tenant to offer the landlord a higher rent and thereby dispossess the original occupier. Quite often, in return for cultivation and manuring, a line of potatoes was grown in a neighbouring farmer's cornfield. Neill estimated that a family of six needed six bolls (approximately 380 kg) of potatoes per year, equivalent to the yield from an eighth of an acre. To cultivate a larger area would require the keeping of a pig to supply enough dung for the ground. Gardens of the lowest quality belonged to farm servants. They had probably undergone no improvement for centuries.

Neill identified several general shortcomings of kitchen gardens (apart from those of the gentry) and suggested ways to overcome them. The greatest need was for longer leases which would encourage occupiers to take more pains. The proprietors could often provide scope for better cultivation by planting hedges to improve shelter, by deliberately siting cottages to minimise exposure and also by having the kitchen garden at the front of the house, rather than in the shade at the back of it. There was great variation in how well the gardens were kept. Too many were infested with weeds, were not dug over during the winter and there was a bad habit of planting the vegetables too closely. Neill recommended that the proprietor should prohibit the keeping of poultry, which attacked the roofs of the houses – evidently of turf or thatch – and damaged the young vegetable plants with their scratching. He much preferred ducks, which would keep down slugs, caterpillars and the like. Given the country wife's attachment to her hens and their produce, such advice would fall on deaf ears. He observed that a cottager with a family often had the use of grass for a cow. Especially in south west Scotland, a pig was highly valued. It could be bought from the proprietor for 10 to 14 shillings and sold for three or four pounds, or the carcase could be salted. Cottage gardens were often too small and should be larger, but not so much as to require the purchase of a pig to supply adequate dung. Fruit trees were often lacking so he recommended a few apple trees like hawthornden, white codling and nonpareil as well as a pear or two, such as green chisel, yair or green achan. Neill recommended that a tenant, on taking up a lease, should be advised by his landlord as to the basic requirements of good management and that this should be recognised as an oral agreement. His colleague Walter Nichol, had been asked by the Board of Agriculture to produce an article on the kitchen garden. This had never been published so he included a copy of it in the appendix, which also included an article by Archibald Gorrie – gardener at Rait – on cottage garden improvement. His final recommendation was for the keeping of bees which could prove a profitable side-line.

He painted a more cheerful picture of the cottage gardens in towns and villages where the occupiers vied with one another to improve their plots. On the first fine day of spring they were to be seen at work on their plots. The eye of the observer was 'gratified in beholding the cheerful countenance and alacrity of the industrious cottagers in regulating their little spots' – economy of expression was hardly Neill's style. On summer evenings they could be seen at ease inspecting the progress of their crops. Who could doubt the value of such a healthy pastime, which discouraged visits to the ale-house?

Although fruit would not be of much concern to farmers, Neill recommended a few hardy apples (as briefly referenced above) – varieties such as hawthornden, royal white, winter codling, nonsuch, leadingtons, ribston and gogar pippins, while, for pears his choice was for carnock, yair, jargonelle, winter auchan, muirfowl egg and green chisel. Gooseberries and currants should be grown – especially the large Dutch white currants which yielded the best sort of homemade wine that he thought might supersede foreign white wine. He could not think of a more cooling summer drink than white currant wine and water.

Neill considered next – in considerable detail – the commercial market gardens which were now to be found about every town or village. He paid special attention to the Edinburgh market and the prices which prevailed. About 1746 a gardener named Henry Prentice was the first to cultivate white peas, potatoes, turnips and several other vegetables for sale in Edinburgh, bringing them in by cart. The volume of produce steadily increased so that by 1812 the total area of ground devoted to market garden produce was 396 acres, tended by 76 gardeners, on plots of five to six acres on average. In some instances, one sixth to a third of the ground was laid down in gooseberries, red and white currants and raspberries. Beyond the immediate precincts of the city, there were about 60 acres distributed between Dalkeith, Musselburgh and Prestonpans, where only part of the produce came to the Edinburgh market, the rest being sold locally. The annual rent of garden ground about Edinburgh was between 8 and 16 pounds sterling per acre. According to his enquiries, unless the soil and aspect were especially favourable, the tenant could barely make ends meet unless the rent exceeded 12 pounds, otherwise the ground was likely to suffer from want of manure and over-cropping. He thought no class of men in the country experienced greater fatigue and exertion than the Edinburgh market gardeners, who in spite of their labours were a habitually cheerful and hearty sort of men, generally of sober habits. A few had done well, lived in genteel style and gave their children the best of educations. He estimated the annual return per acre of well-cultivated ground as between 40 and 45 pounds sterling. The most profitable vegetables included early cabbage, late turnips, early potatoes, onions, leeks, celery, savoys and sometimes carrots.

Escalating demand – due to increase in the population – had encouraged some private estate owners to bring in for sale surplus produce from their own gardens. Neill did not think this a very good idea, not least because such gentlemen were inexperienced in market practice, likely to be taken advantage of and unintentionally sell their produce at less than the going rate, to the discomfort of the professional market gardeners.

The annual value of vegetables sold in the Edinburgh market averaged about £17,000, but with considerable fluctuation between years. The wholesale green market was held on Tuesday, Wednesday and Saturday on the east side of the High Street, opposite the Tron Church, from six to nine in the morning. The wholesale fruit market was held on the same day on the south side of the street. The retail green market had been recently improved to provide 28 stalls, let at nine pence to two shillings and sixpence per stall according to size and situation. The fruit market stalls were set along the west wall of the Tron Church, at a cost of four pence per stall. There was also a market on Wednesday morning for seedling cabbage plants of several kinds, which sold at between eighteen pence to three shillings per thousand.

Neill thought it likely that – apart from Dundee – Edinburgh was better supplied than any other place in the country and certainly boasted the largest strawberry farms in the vicinity of the city. Many of the banks of the north and south Esk a few miles away, were given over to strawberry beds. The fruit was brought into the city in wicker baskets, each containing a Scots pint of fruit, or four English pints. On average 60,000–70,000 Scots pints of strawberries were annually sold in the market. At the beginning of the season, generally about the king's birthday on June 4th, they could sell for the extraordinary price of half a guinea a pint, but later the average was nine pence to a shilling a pint. He estimated the annual value of the crop at between £3,500 and £4,000. On average, an acre of ground produced 800 to 900 Scots pints, with a return of £30 to £40 per acre. Many thousand pints were consumed in the public gardens about Edinburgh and during the season parties were arranged to eat strawberries amid the beautiful scenery of Roslin. The Virginian or scarlet was the most popular variety, although Chile strawberries were also sold.

Since he had described the Edinburgh market in such detail Neill spent less time on other places, but still noted what was of particular interest. He singled out Dalkeith and Musselburgh as well supplied with sale gardens that paid particular attention to fruit of all kinds. This was also true about Haddington. In Fife, every village of any size had a market garden. Kirkcaldy was notable for the amount of fruit grown, especially gooseberries and currants – a substantial amount of which went for winemaking. The great enthusiasm for this domestic enterprise was a direct consequence of the ever-lasting Napoleonic War, the resultant absence of French wines and the patriotic sentiment of replacing them with a homegrown product. Formerly Kirkcaldy had excelled in the production of onions, benefiting from the availability of free seaweed as manure. But latterly, the borough had taken to selling it at two shillings and sixpence a cartload, and the onion crop had declined, casting doubt on the benefits of this innovation.

Dunfermline was rather exceptional in being poorly supplied with sale gardens, but there was less need since almost every house had its own garden. In southern Scotland, Dumfries had about a dozen sale gardens generally about three acres apiece and doing very well. In the southwest there were sale gardens about Ayr and Kirkcudbright, where the favourite crops were onions and carrots. Large quantities of the former were exported to

Ireland. The populous areas of Paisley, Greenock and Port-Glasgow were well supplied with market gardens of variable size. Cabbages, carrots, turnips, early potatoes, savoy and leeks were the preferred crops, together with a great quantity of soft fruit. The ocean going ships which called at Greenock and Port-Glasgow to take on supplies, had resulted in about 60 acres of ground being largely devoted to this trade. One of the largest of such gardens was the two centuries old, six acre garden of Lord Belhaven of Newark. Vegetables taken aboard had to last well – hence potatoes, carrots and onions were in demand, as well as cabbages, which could be stored in a cool place in the ballast. When the market price was high, owners of private gardens brought in produce for sale.

Glasgow – in spite of resembling Edinburgh in population size – was relatively less well supplied with market gardens. The annual value of vegetables sold in Glasgow was estimated at £10,500 and at times reaching £12,000. The items most in demand were early potatoes, early cabbages, early and yellow turnips, onions, leeks, savoys, German greens, Scots kale, celery, late cabbage, broccoli, peas, beans and asparagus. There was hardly any demand for sea kale, artichokes, beet, endive, French beans, garlic, shallots or cauliflower. As in Edinburgh, old-fashioned vegetables such as skirret, scorzonera, cardoon and salsify no longer appeared on the stalls. A few private gardens supplied the rarer vegetables. Neill estimated that the Glasgow vegetable market depended on some 40 'respectable gardeners' who employed about 100 labourers and workmen. The total ground devoted to strawberries rarely exceeded 30 acres. The chief varieties comprised pine, Chile and hautboy. In a good season 800–900 Scots pints were produced.

In central Scotland the recent establishment of a large depot for prisoners of war at Perth had given new life to the market garden business. Onions were particularly profitable. Dundee was as well supplied with vegetables as any other city in Scotland. The produce was of the highest quality. Until the mid-18th Century there were few sale gardens about Dundee but about 1750 a gardener named John Reid rented a near-by piece of ground and started bringing in vegetables for sale on a wheel-barrow, setting a precedent which was quickly followed until there were about 50 gardeners who regularly appeared at the market. Neill estimated the combined area of their gardens at 100–150 acres. All the usual vegetables were sold. Early potatoes, turnips, peas and strawberries were often first on sale on June 4th. The Dundee gardeners vied with one another to produce as many as possible different sorts of produce by that date. Neill noted that salads were better grown and in greater demand here than in Edinburgh. The gardens rented at £10–£14 per acre. Most of the gardeners kept several milk cows which were fed on garden refuse and brewer's 'draff'. Apparently Dundee abounded with small breweries which provided this useful by-product. The gardeners sold the milk, delivered daily to their customers, who paid once a quarter. They were also provided with a steady supply of dung – a good example of 'sustainable' production. As on the west coast, a substantial amount of the produce was sold to shipping.

Further north, about the expanding city of Aberdeen, there had been a great increase of sale gardens during the previous 50 years. Several hundred men were now cultivating ground that had formerly been but a stony waste and was now fetching from £8-£20 an acre. Gooseberries and currants were in great demand. At Banff there were several sale gardens, from one to four acres in size, producing the usual range of vegetables. Neill observed that the market gardeners here were a set of industrious men, who without any other occupation contrived to bring up their families in a very decent manner. Further north, gardens were fewer in this sparsely populated region. Cabbages and coleworts were widely grown in the little private gardens. Onions were often sold in beds, to be lifted and dried by the purchasers who would sell them at about five shillings a peck. At Elgin, the Duke of Gordon owned a market garden which specialised in producing onion that found a ready sale among the fisher folk from Burgh-head to Buckie. At Inverness there were a few small sale gardens, renting at about £6 an acre; onions and soft fruit were most in demand. In the North Highlands there were no sale gardens except for Thurso, where there was a garden of seven acres for which the occupier paid £4 a Scots acre. In the Western Isles, Rothesay in Bute – the only considerable town – lacked a regular market garden, although the people were well supplied with vegetables. Most of them had small vegetable plots which yielded a surplus for sale at reasonable rates. At Stornoway in Lewis there was no market garden, although the need was great. It was an important rendezvous for shipping that could have been supplied with the basic vegetables, as at other ports. Among the Northern Isles there was a complete dearth of sale gardens. Neill concluded – from personal experience no doubt – that there was insufficient demand for vegetables and fruit to sustain a full-time market gardener.

The scarcity of gardens in the Highland areas needs a little qualification. It was nearly 80 years since they were first established in the Highlands. Sir John Cameron of Lochiel was the first to do so, at Achnacary in Lochaber. In August 1734 he entertained his guests with 'hotch-potch' made up of peas, turnip and carrot.[2] Since that time, kitchen gardens were to be found at all the seats of the gentry and their tenants were steadily coming to appreciate the yield of their kale yards.

Neill's survey provides a convincing demonstration that by the early 19th Century, market gardeners and their workforce were engaged in a substantial rural trade in Scotland. The high prices obtained in the Edinburgh market for early cabbage, cauliflower, peas and potatoes would have surprised many a country gentleman and encouraged him to value more highly the produce from his own garden. Neill's account included a comprehensive list of the varieties of vegetable that might be found on sale in commercial gardens. This is included in Appendix 1. The common assertion that vegetables were something of a curiosity in Scotland at that time sits uneasily with this evidence to the contrary.

Orchards

The distinction between public and private sources was much less well defined for orchard fruit than for vegetables and soft fruit. It was widespread practice for large estates to sell their surplus fruit in the local markets. Such private orchards were distributed throughout the country and were especially frequent in south-east Scotland. Places with an early monastic tradition of fruit culture, like Jedburgh and Melrose, had orchards which could be classed as commercial. The planting of fruit trees had of late increased about Jedburgh where there were now about 30 acres of trees, while the banks of the Jed where it flowed through the town, were covered with many fruit trees. There were still ancient pear trees

– remnants of the monastic orchard at Jedburgh – 30–40 feet high with wide trunks and spreading branches. One of them – a red honey – was even between 50 and 60 feet in height. Others now showing their age, were supported by props but still produced good yields. The varieties of these old pears included crawford, achan, longueville, lammas, warden, bon chretien, bergamot, gallert and jargonelle, and of these the warden best withstood the test of time. In a good season a single tree could produce 30–50 bushels of fruit. Neill noted that when some of these old trees died and were dug up, they were found to have been planted over a pavement of flagstones. This ancient practice – which survived into the 19th Century – was designed to promote surface rooting and hence earlier fruiting. Similar old pear trees were also found at Melrose. In 1795 two thorle pear trees had ripened 60,000 pears that had brought eight guineas in the market place – no mean sum.

In Berwickshire the most notable orchard was at Dryburgh. Planted by the Earl of Buchan in 1795 it was very productive. The trees, supplied by Dicksons of Hassendeanburn, were planted 40 feet apart, with the stems protected in straw to keep the hares at bay. It was common practice to grow an undercrop in orchards with a rotation which comprised a green crop of barley or oats, grass for hay, oats and then back to a green crop of barley. Massive amounts of dung were applied. The most popular varieties of apple in the model orchard at Dryburgh included queen of England, hawthornden, white custard, golden custard, strawberry, whorle, fulwood and several kinds of the highly regarded leadingtons. The pears that did best included crawford, drummond, lady lemon, muirfowl egg, green pear of Yair, English bergamot and winter citron. In a footnote, Neill – with his habitual respect for the gentry – acknowledged this information from the 'Noble Proprietor'. Elsewhere, there were two public orchards at Coldstream – one of four acres close to the Tweed, consisting mostly of elderly pear trees but yielding from £150 to £180 per year. The other orchard was still being prepared. The tenant had a lease of eight acres for seven years at £6 per acre and for a further 14 years at £7 per acre, with the obligation to plant the whole as an orchard.

In East Lothian there were no orchards of any size. At Ormiston Hall there were two acres of trees, which though only 20 years old were afflicted with canker. Neill thought that the cold easterly winds of April and May made this district unsuitable for fruit trees. Mid and West Lothian had little to show in the way of orchards either, although a substantial amount of apples and pears were produced in the market gardens for sale in the Edinburgh market. In southern Scotland there were a number of private orchards, some of them a century old in Dumfriesshire. The trees were generally planted much closer together, at 20 feet apart and when let brought about £40 per acre. However, it was in Clydesdale where the largest and most numerous orchards were to be found. Considerably more than 200 acres – chiefly on the banks of the Clyde – were devoted to fruit trees. Often the banks were so steep as to be valueless for arable use. Where level enough, undercrops of clover, potatoes, oats, cabbage and turnip were preferred. The ground was always dug – never ploughed – and heavily manured. Gooseberries and currants were also extensively grown, often yielding a significant proportion of the annual profit from the ground. As the trees matured it was advisable to replace the undercrop with grass and only occasionally dig the ground. Apples were mostly grown although there were plenty of pears and especially plums – which were often grown round the margins of the orchard to provide shelter. Cherries were rarely grown, chiefly because the birds got most of them. The fruit was sold by roup and gathered by the purchaser for sale locally, or even sent to Edinburgh.

The largest orchard was that of Cambusnethen with 25–30 acres, followed by Milton (21 acres), Dalziel (16 acres), Holmfoot (15 acres), Waygateshaw and Brownlee (each of 12 acres). In addition, there were about 16 orchards of two to four acres and many more of smaller size. At Paisley there was an ancient orchard which now found itself in the middle of the town. It formerly belonged to the Semple family and was believed to have been laid down in the time of James I. It also had big old pear trees among which the yellow calder, grey achan and lancel varieties were the most productive, with an average yield of about 500 lb. Neill expressed surprise that Ayrshire – so well suited for orchards – had so few, while Dumbartonshire was no better.

After Clydesdale, the Carse of Gowrie was famous for its more than 20 orchards, lying on the tract of land just north of the Tay between Kinfauns and Dundee. Some of them, like Seggieden, were at least 200 years old while others were between 50 and 100 years. Monorgan was the largest and also one of the oldest. The rent of these orchards depended on the valuation of yield in the autumn. Another area well supplied with fruit trees – although not so many as the Carse of Gowrie – was the Carse of Falkirk, especially in the parishes of Bothkennar and Airth where greengage plums were grown. The soil here was amazingly deep, at least 30 feet and often a great deal more. Aberdeenshire was largely bereft of orchards, apart from those at Monymusk and Skene, which were in a neglected state. The only extensive orchard was at Pitfour, which had been laid down over the remains of the ancient Abbey of Deer. In the West Highlands commercial orchards were absent but there were plenty of small private ones. In Argyllshire several attempts to establish orchards had failed. In Invernessshire every gentleman had a garden with fruit trees but commercial orchards were rare. In the Earl of Moray's garden at Castle Stewart, north of Inverness, there was a collection of morello and kentish cherries and the large black gean. At the Priory of Beauly there was evidence of a former, monastic orchard. Neill suspected that some of the unusual varieties growing in the local cottagers' gardens were descended from the priory fruit. In Rossshire, on the north side of Loch Beauly, there was an old orchard with still productive black and red achan pears. In Sutherland there were only the decayed remains of pear trees, reputed to have been planted at Skibo Castle by the Bishop of Caithness. In the Western Isles apples and pears were grown in several small orchards in Bute. On the island of Coll there was an excellent kitchen garden with a good collection of apple, pear, plum and cherry trees. On Lismore the Roman Catholic Seminary had established a productive garden and orchard. By contrast, in the Northern Isles orchards were unknown apart from a few apples grown on walls in Orkney.

Generally apples, pears and plums were grown in Scottish orchards, occasionally cherries, with gooseberries and currants as a common undercrop. Neither cider nor perry was made in Scotland, nor was it likely that these wholesome beverages would 'soon be able to supplant pernicious whisky, which affords a much stronger stimulus and is through habit a favourite with the common people'. The markets of Edinburgh and Glasgow and the manufacturing towns in the west of Scotland offered a ready market for the entire orchard fruit produced in the country, and even for many cargoes from England and America. The brisk demand for gooseberries and currants depended on the domestic production of wine and also the making of vinegar. As much as £100 had been paid for an acre of gooseberries about large towns. Neill thought that the better kinds of plum were not suited for orchards – unless the market was close by – since they did not travel well. Separate cherry orchards were unknown.

Most orchard owners did not let them by the acre or on lease – except occasionally in the case of new or young orchards – but sold the crop at the end of the season, often retaining the undercrop for their own use or let it separately. In England, apple trees were often pruned up the stem to produce standards that meant the older branches were out of reach of cattle, whereas in Scotland such pruning was rare, since stock were not usually turned into orchards. In private orchards, two or three cows might be pastured in the orchard and provided with a loop of rope round neck and fore-leg to allow browsing but prevent damage to lower branches. Few of the varieties of apple and pear which were favourites in England were grown in Scotland, except for the golden and ribston pippins and a few others. Even fewer of the pears preferred in England were to be found in Scotland. The langland pear, which was the general favourite of the English farmer, was unknown north of the Border.

It was a general rule among Scottish orchardists that half of the apple trees should be harvest and winter fruit, which kept well and could be sold at market throughout the year.

It was recognised good practice to plant a considerable number of different varieties, of both early and late kinds, to increase the likelihood that some would escape the unpredictable hazard of spring frosts. Neill observed that the laying down of an orchard was something of a lottery since only experience would identify the best variety of apple or pear for a particular soil and aspect. Neill concluded his account with a well informed technical summary of orchard management and the diseases of fruit trees, which we can pass over.

His report was supplied with an extensive appendix that included observations contributed by different authors, including lists of the orchard fruit commonly grown in Clydesdale and the Carse of Gowrie, as well as his own comprehensive and historically-valuable list of the varieties of orchard and soft fruit grown in Scotland. He enumerated 101 kinds of apple, 65 pears, 28 plums, 14 cherries, eight apricots, eight peaches, four nectarines, three figs, 25 gooseberries, five currants and seven raspberries. In favoured situations – usually but not invariably with the benefit of a hot wall – apricots, peaches and nectarines were widely grown on the large estates. He added an informative comment about each variety. A few examples are worth quoting to illustrate the kind of information he provided. Thus, for apples, we have:

> **Nonpareil**. Requires a wall, and a south aspect, in a sheltered situation to bring to maturity. It keeps very long, and is one of the best russets for the table.
>
> **Golden Pippin**. South wall. Excellent for the table, from November to March, and keeps tolerably well when pulled in a ripe state. The tree is extremely apt to canker in Scotland, particularly if the soil is wet.

Ribston Pippin. 'A universal apple', says Mr Nicol: 'for these kingdoms. It will thrive and even ripen at John-o-Groat's, while it deserves a place at Exeter and Cork.' It is very fit for the table, and for the kitchen is unrivalled. It keeps long and bears abundantly. Wall or espalier.

Grey Leadington. This is the best of the leadington family. It grows freely, bears pretty well, and the fruit keeps long. Dr Gibson thinks it 'superior to every other apple, if we except the golden pippin and golden rennet; nor (he adds) is it excelled "by either of these kinds".[3] It is named from Leadington or Leathington, the ancient seat of Secretary Maitland, near Haddington. Standard or espalier.

Lemon Pippin. A very pretty yellow and somewhat russety apple of large size. The tree very soon begins to bear. It is a Clydesdale kind, but is cultivated as north as Inverness. This is one of the best apples for drying in a slow oven. Standard or espalier.

Pears: Grey Achan, Black Achan or Winter Achan. A capital late or winter pear, said to be of Scottish origin. Wall or espalier.

Muirfowl Egg. Often placed against walls in Scotland, but the standard fruit much higher flavoured. It is a well known autumn pear and keeps well. It is said to be originally Scottish.

Crawford or Lammas pear. It is one of the earliest pears in Scotland, and the tree is an excellent bearer. It is very generally spread over the country as a standard.

Longueville. A good summer pear, and keeps for some weeks; the longueville of Jedburgh seems to be a variety; it keeps for months. Standard.

Briar-bush or Nettie. Of Clydesdale origin. It is a small but good winter pear, ripens even in indifferent seasons, and the tree is generally a profuse bearer as standard.

Plums: Green-gage. 'The best' says Mr Nicol: 'the most generally known and most highly esteemed of the plum kind.' In Clydesdale and the Carse of Gowrie, and some parts of the west coast, near Greenock, and about Bothkennar in Stirlingshire, where the soil is a rich, deep, hazely loam with a dry bottom, it ripens in tolerable seasons on standards and acquires a higher flavour than when fostered with a wall.

There are several varieties of gage; particularly the yellow or white, and blue or red gage. Neither of these is so common in Scotland as the green, nor are they ever to be seen as standards.

White Magnum Bonum, Yellow, White, Holland Magnum or Egg-plum. This is very common; the tree grows freely and seldom fails to bear, either on a wall or as a standard.

Red Magnum Bonum or Imperial. This is also a large fruit; but chiefly used for baking and preserving. The tree is a very great bearer as a standard.

White Corn, Summer Scarlet, Winter Scarlet, Damsons, Julians, Burnets, Horse-jags are all propagated by suckers, and thrive and bear without any trouble. All of these are common in many places in Scotland, being often planted in hedge-rows, particularly in Cydesdale, Carse of Gowrie, about Melrose and Kelso and several places in Dumfries and Ayrshire. They are excellent trees for cottage-gardens, requiring little care, but affording abundance of fruit for pies.

Cherries: Mayduke. 'We have no cherry' says Mr Nicol, 'equal to this. It thrives in all situations'. It does very well as a standard, but on a good wall and with a southern aspect, the fruit becomes considerably larger, and higher flavoured.

Kentish. This is the kind generally planted in orchards and market gardens. As the flowers are late expanding, they generally escape the spring frosts, and afford a plentiful crop. The fruit however, is fit only for tarts. The tree answers well either as standard or buzelar.

Morella. Though the taste of this cherry is not agreeable to many, yet when ripened on a wall in the full sun, it acquires a size and richness of flavour superior to any other. The tree grows freely and bears well; it is often placed on a north wall.

Neill's catalogue is also useful in identifying fruits believed to be of Scottish origin, or at least long association, since he often uses names mentioned in the earliest, late 17th Century Scottish records[4] but not referred to by later authors. The 'Nicol' whom he quotes was Walter Nicol, the prominent horticulturist and writer who was briefly fellow secretary of the Caledonian Horticultural Society until he sadly died prematurely. Dr Gibson was the 18th Century (1768) writer on the varieties and management of fruit trees.

Other items in the appendix referred to reports on orchard management, a case for growing early potatoes and also a notice 'concerning the taste for flowers among the Operative Manufacturers at Paisley, and Account of the Florists Society instituted there'.

This casts the good folk of Paisley in such favourable terms that it is worth quoting, although Neill's literary style is unduly long-winded. 'The operatives at Paisley, taking them at large, exhibit a condition of improvement very rarely indeed if at all to be paralleled among persons in the same rank of life; and they are particularly remarkable in their taste for objects which please the eye by their beauty, for such occupations for amusement as require nice attentions, and for various intellectual gratifications. In their dress, in the furnishing of their houses, and in matters of a similar kind, they study a degree of neatness. Even their pigeons, which they keep in considerable number, are known in the vicinity to be distinguished for their beauty and variety. Several operatives greatly excel in the management of bees; and communicate to each other their experiments and success. It will perhaps be difficult to find elsewhere in the same classes of the community, an equal proportion of persons who occasionally entertain themselves with *making verses*. And it is probable, that for miscellaneous information, they are not to be equalled by operatives in any place.' In a footnote, he stated that the only operatives in this nation – or perhaps in any other – who can be compared with them for information, are the miners at Leadhills, 'who work but six hours a day, and have with success devoted much of their leisure to the improvement of the mind. But the intellectual attainments of the Paisley operatives must be more various than theirs, and are probably in no department inferior.'

He continues: 'Some idea of the zeal of the Paisley cultivators may be formed when it is stated, that several of them can show 70 or 80 of the most choice variety of pinks; others 200 varieties of the choicest tulips; others 60 or 70 varieties of the best carnations, all named; besides many excellent inferior varieties of each kind.'

He then went on to 'advert briefly what are reckoned by Paisley florists the characteristics of a good pink; and the best methods of bringing this flower to perfection. Those flowers which are well laced, and least notched in the edge of the petal, and, as nearly as possible, rose-leaved, are considered as the grandest.' He resumes with more details of the particular combinations of colour and petal shape which were prized, ending with the acknowledgement that this information about the Paisley florists was imparted to him by John Findlay, gunsmith in Paisley – 'a very intelligent and successful florist'.

Neill was at pains to make his account of 'Scottish Gardens and Orchards' as complete and up-to-date as possible and any historian of the Scottish rural scene must be grateful for the trouble he took.

1   Patrick Neill, *On Scottish Gardens and Orchards*, By Desire of the Board of Agriculture, 1813.
2   Elizabeth S. Haldane, *Scots Gardens in Old Times (1200–1800)*, 1934, p.7. Alexander Maclehose & Co.
3   John Gibson, *The Fruit Gardener*, 1768, J. Nourse, London.
4   The Earl of Crawford, *List of Fruit Trees*, Saltoun Papers, MS, 1749, ff.6–19, The National Library of Scotland.

Appendix 1

Neill's list of the varieties of vegetable on sale in Scottish market gardens. Asterisks indicate the varieties most commonly grown in farmers' gardens.

Cabbages: *early dwarf, *York, *drum or Scotch may, imperial, sugar-loaf, red borecoles or German greens; *green and *red curlies and *Scotch kale.

Savoys: *green; globe and yellow curled.

Cauliflower: early and late.

Broccoli: different varieties of *early purple; *hardy green; dwarf purple; early green; dwarf sulphur; white and cape broccoli which had just been introduced.

Carrots: *early horn; *orange; large and red.

Parsnip: common.

Turnips: *early Dutch; *Dutch yellow; *large yellow; and stone. French turnips were raised in a few gardens and Swedish turnip or ruta-baga.

Potatoes: *ash-leaved; early dwarf and many varieties with local names only.

Peas: *blue prussian; *charleton; *dwarf marrowfat; *royal; early frame; early hotspur; golden hotspur; tall marrowfat; dwarf prolific; Leedsmans'; Spanish; sugar; rouncival; pearl; egg; Spanish morato and tall sugar with edible pods.

Beans: *early mazagan; *long-pod; *windsor; sword long-pod; white-blossomed; toker; sandwich; dwarf fan and green Genoese.

Onions: strasburgh; *deptford; *reading; *red; silver skinned; James's; Portugal; welch; Spanish and globe.

Leeks: *true Scotch and *flag-leek.

Chives or cives: *common.

Spinage: round-leaved and prickly.

Mountain spinage or orache.

Parsley: *curled and smooth.

Artichokes: red and white.

Asparagus: red.

Beet: red and green.

White beet and the large tooted variety called mangelwurzel.

French or kidney bean: early yellow dwarf; red speckled; black speckled; Canterbury dwarf; Chinese dwarf; white Dutch; negro; liver and white and scarlet runner.

Celery: upright solid; red; Italian; north's and turnip rooted or celeriac.

Lettuce: *hardy green cabbage; Dutch brown; white; green; Egyptian; brown and spotted cos; Hammersmith; imperial; tennis-ball; grand admiral and Cilician.

Radish: *scarlet; short-top; early frame; white and red turnip-rooted; black Spanish;, long white and salmon.

There were also a variety of herbs which included *spearmint, *peppermint, *thyme, horseradish, balm, red and green sage, plain and curled tansy, costmary, *cresses (plain, curled and broad-leaved), Indian cress or nasturtium, white mustard, common and French sorrel and camomile.

In what Neill termed 'a superior class of garden', there were generally also endive, Brussels sprouts, sea-kale (introduced since 1805), Jerusalem artichoke, Hamburgh parsley (i.e. a variety with large roots), garlic, shallots, rocambole, patience-dock, green and golden purslane, both winter and summer savory, fennel, small burnet, sometimes chervil, angelica, lovage, lavender, hyssop, tarragon, rosemary, borage, marigold, common clary (generally for clary wine), caraway and sometimes coriander for their seeds, sweet marjoram and pot marjoram, sweet basil and bush basil for a warm border or in front of a hot-house.

Others, which were raised in a frame and later planted out in a warm border, included:

Capsicums: long red; long yellow; ox-heart; cherry and cayenne.

Cucumbers: short for pickling.

Tomato or Love Apple: both red and yellow.

There were many varieties of pumpkins and gourds as well as melons kept in frames and mushrooms grown in a dry bed or an out-house.

7   *Botanist and Plantsman*

Neill's fascination with all aspects of natural history was dominated by his love of plants, both wild and cultivated. This was manifest in his botanical excursions to collect wild plants, his garden and his unflagging devotion to the cause of horticulture. The Caledonian Society's successful growth and influence among estate owners, head gardeners and professional nurserymen owed a great deal to Neill's sustained enthusiasm and conscientious administration. By the early 19th Century he was recognised as an authority on gardening and horticulture. His report on Scottish gardens (1813) and his articles on horticulture featured in the *Encyclopaedia Britannica*[1] and Brewster's *Edinburgh Encyclopaedia* respectively.[2] Together with two colleagues he was commissioned by the Caledonian Horticultural Society to examine the state of horticulture in northern Europe in 1817, followed by a private visit in 1821 to supplement the survey. These visits are described in the next chapter. His closest friends and correspondents included the leading botanists of the day – namely William and Joseph Hooker, Robert Brown, Robert Greville, William Arnott, David Don, James Mackay and his brother James Townsend Mackay – as well as the successive professors of botany at Edinburgh University – Robert Graham and John Hutton Balfour. Neill was a particularly close friend of both the McNabs – father and son – who, in succession, supervised the Royal Edinburgh Botanic Garden for so long. He was a welcome visitor to the premises of all the local nurserymen who were eager to show him their latest novelties and acquisitions. Through his duties at the Caledonian Horticultural Society he won the friendship and respect of the head gardeners throughout the Lothians and beyond. As a young man he went on botanical excursions in the Scottish countryside to collect wild plants for his herbarium, which was later expanded to include specimens of exotic species, often obtained from the Botanical Garden. He was familiar with every aspect of horticulture in Scotland and was widely respected for his scholarly appreciation of its finer points and the rapport he enjoyed with so many working gardeners. In a different context he supervised the design and planting of East Princes Street Gardens and gave advice on other garden sites in the city.

    Canonmills Cottage had been the Neill family home for about a century. Situated about a mile from the centre of Edinburgh on the north shore of Canonmills Loch, it was close by the mills that had formerly belonged to the canons of Holyrood House. Although only a little more than half an acre in extent, the garden contained a remarkably diverse collection of plants of all kinds. Shrubs, trees and climbers were set about the borders to provide attractive views from the house. There was a pond for water plants. The walls were covered in creepers that according to Audubon – who got to know Neill in 1826 – provided a home for 'thousands' of sparrows. There was a bank for alpine plants, both a cool and warm greenhouse and a hot-house in which he grew a wide range of uncommon or rare exotics – quite a number of which had been grown and flowered in Scotland for the first time in his garden. In addition, there was a double pit for plants needing protection outdoors and two large frames. These various enclosures housed more than 2000 pots which required regular watering.[3] It was indeed a mini-botanic garden, independently valued at £600 – a very considerable sum in those days. No less an authority than Loudon considered it the best-endowed suburban garden in the country.

The only way to demonstrate the extraordinary diversity of Neill's collection is to list the genera, represented by one or more species, which were often rare or uncommon. The following list, based on Loudon's[4] account, refers to plants grown between 1826 and 1836. A substantial number of these – described in *The Edinburgh New Philosophical Journal* by Professor Graham or Hooker's *Exotic Flora* – flowered for the first time in Britain in Neill's collection and many of them were illustrated in Curtis' *Botanical Magazine*. The number in brackets in the following list refers to representation by more than one species. Occasionally, as part of on-going taxonomic revision, the genus name has since changed but this has been ignored since it is the extent and degree of diversity of the collection which is our concern.

The list of genera is as follows: *Adenophora, Agrostemma, Alstromeia (8), Andromeda (3), Anthyllis, Arbutus (2), Arthrostemma, Astragalus, Aubrieta, Begonia (3), Bellium (2), Bellevalia, Brachystelma, Brexia, Calandrinia, Calceolaria, Calliopsis, Cologania, Cypripedium (8), Cytisus, Daboecia, Dodecathon, Ensellnia (14), Epigaea, Eryngium, Eucrosia, Euphorbia (2), Fernanderia, Ferraria, Francoa (2), Fritillaria (2), Gaillardia, Galanthus, Gardoquia, Gentiana, Gloriosa, Gratiola, Gypsophila, Houstonia (3), Hunnemannia, Isopyrum, Jatroplia, Kennedia, Leontice, Lilium (2), Lobelia, Lupinus (2), Lychnis, Lycopodium, Mahonia (3), Manettia, Mantisia, Milla, Nierembergia (5), Lophospermum, Nicotiana, Nutallia (2), Ornithogalum (2), Oxalis (22), Parnassia (3), Passiflora, Phlox, Phyllodoce, Physianthus, Portulaca (2), Primula (29), Potentilla (3), Ranunculus (14), Rosa, Rulingia, Ruta, Sarracenia, Saxifraga, Schizaanthus, Scutellaria (2), Silene, Sisyrinchium (2), Stylidium, Teucrium, Thalictrum, Tropaeolum (2), Viola (3), Wedelia, Xylobium, Zephranthes, Zigadenus.*

In addition to these herbaceous plants, the following trees and shrubs were to be found in the borders: *Abies clambrasiliana, Abies douglasi, Abies tortosa, Ailanthus glandulosa, Aloysia citriodora, Araucaria imbricata, Araucaria brasiliana, Catalpa springaefolia, Edwardsia grandiflora, Erythrinia laurifolia, Maclura auriantica, Phormium tenax, Pinus bnksiana, Pinus carmanica, Pinus laricio, Pinus pinea, Pimus ponderosa, Piptanthus nepalensis, Pittoporum tobra, Pseutosuga menziesii, Rhododendron arboreum, Ribes sanguineum, Ribes speciosum, Tropaeolum pentaphyllum, Veronica decussata, Yucca gloriosa.* Loudon noted that many of these were well grown specimens.

Drawing of Patrick Neill's garden at Canonmills, Gardeners Magazine, Vol. 12, 1836

On the walls there were *Acacia vertcillicata, Acacia deallata, Cistus ladaniferus, Escallonia rubra, Grevillea rosmarinifolia, Mespilus japonica, Myrtus commnis* var. *latifolia, Passiflora coerulea* and *Wisteria cosequiana*.

The stove contained quite a collection of orchids belonging to the genera: *Bletia, Bletilla, Brassia, Calanthe, Catasetum, Dendrobium, Epidendron, Habenaria, Oncidium, Renanthera, Stanhopea, Vanda* and *Xylobium*. In addition, the following non-orchidaceus genera were represented: *Brexia, Brugansia, Ceropegia, Cinnanmomum* (3), *Dillenia, Dionaea, Euphorbia, Nepenthes* (3), *Russelia, Sida, Solandra, Sinningia, Swietenia, Tacasonia* and *Vellezia*.

It is obvious from this impressive list that Neill's approach was that of a botanical collector. His collection was so varied and with so many newly introduced species that his high standing among professional botanists is not surprising. Naturally one wonders how he managed to gather so many newly introduced and/or rare species. Neill was in regular contact with a number of collectors who were exploring the flora of distant parts. Thus he benefited greatly from his friendship with John Gillies – an Edinburgh naval surgeon – who went to Buenos Aires in 1823 and lived there for several years at Mendoza before returning to Edinburgh in 1829. Gillies sent him seeds and tubers of South American species. John Tweedie – a Scots gardener – also went to Buenos Aires and sent Neill the products of his collecting. These two collectors account for the many South American species in his collection. Neill contributed to the collecting expeditions of the Drummond brothers Thomas and James, who supplied him with material from respectively North America and Australia. We know he received seeds from a collector in St Petersburg and from a relative – Robert Neill – in government service in Tasmania. No doubt there were others, whose names have gone unrecorded. He was on excellent terms with the successive professors of botany at Edinburgh University and the McNabs. He and the McNabs often shared newly imported seeds and there was a friendly rivalry in seeing who could first bring them into flower.

No one played a more important role in his botanical interests than William Hooker whom he got to know when Hooker was Professor of Botany at Glasgow University. Their acquaintance developed into a warm friendship that never faltered and remained strong after Hooker moved to Kew. From before 1830 to the time of his death Neill corresponded regularly with William Hooker, often sending him recently introduced plants that he had flowered and receiving specimens of rare species in return.[5] For example on July 19 1830 when Hooker was still at Glasgow, Neill sent him a vasculum containing part of a specimen of a South American species he called *Alstromeria polita*, with a drawing by Neill's friend Dr Greville, who as well as being an expert on algae, mosses and liverworts, was also a capable artist. Neill also included a branch of a *Portulaca*, grown from seed he had received from Gillies, and enquired whether Hooker would like a specimen of *Verbena intermedia* and also whether he should ask Greville to sketch it for him.

About the same time Neill informed Hooker that he had heard from Professor Graham that Thomas Drummond was going to collect plants along the west coast of North America. Although he did not want dried specimens, he would like seeds or bulbs of any rare or new species and so would like to add his name to the list of subscribers. He wished to advance

£5 with the prospect of increasing it to £15 if it seemed justified. In early January of the following year he informed Hooker that he delayed sending a specimen of *Lophospermum* (probably *Asarina lophospermum*) until the intense frost had abated. He had sent off by the Glasgow wagon a few plants for the Glasgow Botanic Garden addressed to Mr Murray (the head gardener), together with a small plant of the *Lophospermum* for Hooker, adding that he would always be glad to send a rare plant to him or his friends. Hooker had written to a Mr Parker about a parcel of material from Buenos Aires, possibly collected by Tweedie. One of the tubers in the last packet looked as if it would turn out to be a *Tropaeolum* – probably the climber *T. pentaphyllum* – which appears to have flowered for the first time in Neill's garden and was illustrated by Hooker in his *Exotic Flora*.

In April 1832 Neill acknowledged receipt of two packets of seed from Drummond – one addressed to himself and the other to Mr Lawson, the nurseryman. The seed packets were numbered but not named. He had sown some of them but they turned out to be reedy grasses or umbelliferous species which were too big for his garden. At Hooker's request he was sending a portion of the seeds to the College Botanic Garden at Dublin. He remarked that he had a *Pontidera* from Buenos Aires in fine flower. What he termed the saccharine *Lippia* was also in flower and both a new *Millardia* and a *Francoa* were in bud. Two species of *Acacia* probably both from eastern Australia – *A. verticillata and A. mollissima* – had flowered that spring in the open border of his garden.

In September of the same year Neill wrote to Hooker that his late gardener Alexander Scott, who had moved to London, had been asked by Mr Knight of the Exotic Nursery at Chelsea, how best to subscribe to Drummond's American expedition. Later that year Neill expressed regret that Drummond's seeds had proved disappointing since they had produced gigantic, umbelliferous plants, quite unsuited for his garden. So far the only desirable item had been a root of *Lawrencia psittacum*. In October 1846 he informed Hooker that a *Brugansia sanguinea* and a *Stapelia insignis* had flowered in his greenhouse.

In 1838 he wrote that he had received seed from his relative Robert Neill in Tasmania and also from Professor David Don from Nubia. In 1844 he reported receipt of 50 packets of Australian seed from Thomas Drummond, but there were only 35 different sorts and there were many duplicates or triplicates. Apparently Hooker had sent the packets to him. Neill was well aware Hooker was a busy man, but could he identify seven of the numbered samples? In February 1842 Neill enquired of Hooker whether he had had any news of two ships bound for American exploration. He had heard a rumour that one of them was wrecked on a coral reef. A former gardener of his – William Brackenridge – was on one of them and he was therefore greatly concerned. However it later became known that no harm had come to Brackenridge, who stayed on in America to enjoy a successful horticultural career. He became superintendent of the National Botanic Garden in Washington and later ran a nursery in Maryland.

In July 1844, Neill again referred to his relative Robert Neill – who was a 'capital draughtsman' – and had taken Drummond (who spent several days with him) to see an *Acacia* which Robert Neill had drawn and which Drummond thought was new. Later the same year, he thanked Hooker for cuttings of *Sedum ochroleucum* and also for seeds of a grass from the Falkland Islands. In February 1846, he thanked Hooker for the present of live plants of *Cephalotus follicularis* (an Australian pitcher plant), and also of *Phytolephas macrocarpa*, probably from the same region. In 1847 he reported receipt of six seeds of Drummond's fine new *Hakea*, apparently sent by his relative Robert. They had germinated and, if they survived the winter, he would send Hooker one.

It is important to note that Neill had the good fortune to employ a succession of very able gardeners, who often went on to important jobs in their profession. They were very skilful men who so often brought into flower species that they had never met before.

The other person with whom Neill regularly corresponded about plants was Walter Calverly Trevelyan. He was a man of great wealth who became Sir Walter when his father died. He lived the life of a country gentleman, travelling leisurely about Britain and Europe with his wife, in pursuit of his scientific interests. Although better known as a geologist he appeared equally interested in botany, collecting plants for his herbarium as well as specimens and seeds for his botanical friends, who included Neill.

The substantial number of surviving letters from Neill to Trevelyan deal almost exclusively with wild species from Britain or Europe, which Trevelyan had found on his excursions.[6] These letters are particularly significant since they reveal Neill – as field botanist as well as gardener – just as keen to receive seeds for his collection of some rare British or foreign plant, of no ornamental value, as exotic specimens from the tropics.

In April 1834 Neill reported that the *Armeria* seeds sent to him by Trevelyan from the Channel Islands and the seeds of *Trichonema bulbocodium* (*Romulea columnae*, the sand crocus) from Germany, had germinated very well. Apparently Trevelyan had also found this rare *Romulea* in east Cornwall near the mouth of the river Exe. Neill had already told McNab about this interesting find and would pass on the news to Professor Graham. James McNab – William's son – was anxious to obtain two or three dried specimens, together with dissections of the ovary. He was about to set out on a plant hunting expedition to North America, hoping to cover his expenses by providing dried or live specimens. If any one wanted to subscribe there was no time to lose in doing so. Neill then added a comment about the time he visited Oxford and delighted in 'filching' a patch of *Sedum dasyphyllum* (an introduced species of stonecrop) from the top of a wall as well as seeds of *Senecio squalidus* (the Oxford ragwort), both of which flourished in his garden but had later disappeared. *Ribes speciosum* was flowering with him for the first time in Scotland, while in the Botanic Garden the *Diapensia lapponica* was in beautiful flower.

On April 23 1839 Neill hoped Trevelyan and his wife would visit Canonmills, where they would find most of the plants from the south of France in good health. No doubt this refers to an earlier consignment from Trevelyan.

On January 10 1843 Neill wrote to Trevelyan thanking him for the present of seeds he had recently collected in Greece. He had asked McNab to divide them into three packets: one for Professor Graham and the Botanic Garden, one for the Experimental Garden and the third for himself. They were all very grateful for his 'obliging attention'.

In July of the same year Neill wrote again on botanical topics, especially plants he had received from Trevelyan. The root of *Plumbago lanceolata* with monstrous spikes had been potted and sunk in a slight hotbed. He possessed a similar variety of the related species *P. major*. The specimen of *Lathyrus maritimus* (*L. japonicum*, a local coastal species), both in flower and pod, was 'capital'. McNab, who happened to be with him, got a pod and a bit of root and Neill hoped the bits of root he had planted would take. Another specimen was definitely *Lotus angustissimus*, but being annual, he would not be able to keep that going. Trevelyan had offered to send him seeds of native species from Suffolk and this moved Neill to comment on East Anglian species he had never seen. He had for many years longed to see the *Lathyrus maritimus* in its native habitat but he would now have to be satisfied with Trevelyan's description of it. Just in case the plant he had received should fail, could he have two ripe pods? He would find a ripe head of *Trifolium ochroleucon* (the sulphur clover) very desirable, since he did not have it in his collection. He could think of several Suffolk species which would be highly prized at Canonmills but could not say whether they were to be found about Southwold where Trevelyan was staying. He had never seen *Sagina maritima* (sea pearlwort). He had longed for *Holosteum umbellatum* (jagged chickweed, now extinct) It was said to grow on thatched roofs. Other Suffolk plants he would love to have included: *Hypericum elodes* (marsh St John's wort), *Genista pilosa* (hairy greenwood), *Vicia lutea* (yellow vetch), *Lathyrus aphaca* (yellow vetchling), *Tillaea muscosa* (*Crassula tillaea*, mossy stonecrop), *Bupleurum tenuifolium* (*B. tenuissimum*, slender hare's ear), *Galium parisiense* (wall bedstraw), *Dipsacus pilosus* (small teasel), *Chlora perfoliata* (*Blackstonia perfoliata*, yellow wort) and *Spartina striata* (*S. maritima*, small cord-grass). Trevelyan had sent him seeds of a grass which was possibly *Poa rigida* which was also to be found on Salisbury Crags in Edinburgh. He had the impression *Fritillaria meleagris* was recorded from Suffolk. He would gladly try seeds of any of the *Orobanche* (broomrape tribe) as well as English orchids. To round off his long list of 'desiderata', if Trevelyan happened to come across *Bryonia dioica* (white bryony) or *Tamus communis* (black bryony) in seed he would very much like a little of each of them. This letter perfectly illustrates Neill's keen interest in the native flora. He was happy to grow inconspicuous species, valued for their scarcity or from a particular habitat, and seemed as intrigued by them as much as the exotic inhabitants of his greenhouse and stove. He ended his letter with news that he had received a head of tussock grass from the Falkland Isles.

In July of the same year Neill continued his report on plants received from Trevelyan. He had five pots of sand covered with *Tillaea muscosa* and *Trifolium suffocatum* and wished there were a meeting of the Botanical Society so he could 'regale the eyes of our keen collectors with a sight of living plants'. He then added to his 'desiderata'. A few ripe capsules of *Silene anglica* (*S. gallica*, small flowered catchfly), as well as seeds of *Orobanche minor* (common broomrape), *Centaurea calcitropa* (red star-thistle) and *Torilis infesta* (*Caucalis sp.* hedge-parsley) would all be very welcome. In July 1846 he thanked Trevelyan for another packet of seeds which he shared with Professor Balfour.

There is abundant evidence that Neill established an extensive herbarium of dried plants. He donated duplicates from his collection to the Edinburgh Botanical society at its inception. He also presented between 3000 and 4000 specimens to the Edinburgh Botanic Garden in 1841.[7] The hope that the records of their origin might indicate where he had collected plants in Scotland proved fruitless since the collection has been dispersed or perhaps partially discarded, so that it is no longer possible to identify the original specimens. There is however reliable evidence of Neill's collecting. Nelson has drawn attention to William Ramsay McNab's herbarium (preserved in the National Botanic Garden, Glasnevin, Dublin) containing a substantial number of specimens of Scottish garden plants that were labelled as coming from Neill's collection.[8] The Irish collection was started in 1805 by Neill's friend William McNab – William Ramsay McNab's grandfather. William's son James added North American species as well as other species collected in different parts of the world by other collectors. Neill's plants – which comprise 225 different species or varieties – are dated between 1779 and 1805 and the first letters of the genus name only run from A to M – both indicating the very incomplete record of Neill's contribution to the herbarium, which includes both native and exotic garden plants. Of the total of Neill's herbarium plants, just fewer than 54 percent originated in the Edinburgh Botanic Garden. A few specimens are from his Canonnmills Garden, from Dickson's nursery, from the Dalkeith or Carlourie estates or the remnants of Dr Fothergill's garden, while the rest are unallocated. One specimen, *Antirrhinum speciosissimum*, dated 1805 from the Edinburgh Botanic Garden, had a romantic origin. He writes: 'In the year 1799 a French packet from Egypt being chased by one of our ships of war threw her despatches overboard. A British sailor native of the town of Dundee leapt into the sea & saved them. Among them was a parcel of seeds, from which this elegant antirrhinum sprung.'

As previously noted, Neill was a keen field botanist in his younger days. Sadly the dispersal of his herbarium and loss of his journals prevent us from discovering where he went and what he found – information which would probably have won him an honourable place among Scottish field botanists. All we have are a few desultory references to native plants gleaned from a few surviving letters to his botanical friends – just a flavour of his plant hunting. Thus, in a letter of 1804 to J.E. Smith, the contemporary authority on the British flora, he commented on the limited extent of the flora of the islands.[9] However, he had brought back specimens of 205 species and distributed among 141 genera, excluding chickweeds and nettles. When noting particular species the Latin name Neill used has often changed. In such cases the current name is shown in brackets. Neill listed a few of his rarities, including *Primula farinosa*, first shown to him on the island of Sanday by a Miss Alcock, and later found by him in abundance on Westray and Pomona. He found Hoy a rewarding place for a botanist, finding for example: *Hypericum elodes* or marsh St John's wort; *Andromeda alpinum* (*Arctous alpinus*), black bearberry; *Salix arbuscula* or mountain willow and several species of clubmoss (*Lycopodium*). On Rousay he found *Epilobium angustifolium* or rose bay willowherb, which had not then acquired its current ubiquity. He noted that the Standing Stones of Stenness were covered with what he called *Lichen calicoris*. He found two species of plant in Shetland that he did not encounter in Orkney, namely a small hairy form of *Jasione montana* or sheep's bit and *Scilla verna* or Spring squill. He had picked up a few Orkney names of local plants such as 'yule grass' for meadowsweet, 'arby' for thrift, 'eccle grass' for butterwort, 'pounce' for Yorkshire fog and other soft grasses and 'fibla' for reeds or any tall grass.

In 1808 he informed Smith that he had advised Robert Maugham and his son (Maugham was a founder member of the Edinburgh Botanical Society) as to where to find masterwort *Imperatoria* (*Peucedanum ostruthium*) in the Ochil Hills. Maugham senior thought it indigenous but it is now regarded as introduced.[10] In a further letter in 1811 he remarked that he had travelled from Arbroath to Perth in the company of Robert Brown – a Perth nurseryman – who took him to a new locality of *Linnaea borealis*, the twinflower.[11] He enclosed a couple of specimens, one of which he thought might be *Andromeda* (*Phyllodoce*) *caerulea*. If that were so, it would have been a remarkable find since it is so rare. He returned via Aberdeen where he hoped to meet Professor Beattie but was shocked to arrive in time for his funeral. His friend Brown made regular trips to the mountains in search of rarities like *Gentiana nivalis* or snow gentian and *Bartsia alpina* or Alpine bartsia. Describing a recent visit to the Western Isles in 1810, Neill noted that Macdonald's Cave in Skye was decorated with masses of *Rhodiola rosea* or roseroot and hart's tongue fern. He found the hills behind Greenock covered with *Carum vericillatum* or whorled caraway, just as Mackay had told him many years ago. In Arran he found *Hypericum androsaemum* or tutsan, as Lightfoot had noted. Purple loosestrife (*Lythrum salicaria*) was the commonest weed in the moist oatfields. The rocks of the little island of Pladda were covered with *Cotyledon umbilicus* or navelwort. Neill's month-long trip in the Western Isles, round the north of Scotland and various sites of geological interest was a most pleasurable experience. The weather was perfect and he had the pleasure of the company of his friends Dr Barclay (the anatomist), Robert Stephenson (the lighthouse engineer of Bell Rock fame) and a Mr Oliphant. In yet another letter to Smith in 1814, Neill mentioned that a specimen of *Solidago lanceolata* (*graminifolia*) – a goldenrod – had been brought to him from Inveresk near Edinburgh, in 1800 by his cousin Adam Neill.[12] He had correctly concluded that it was American and had grown it in his garden.

We can infer that the rarer items in Neill's plant collection of both live and preserved specimens were mostly derived from collectors working abroad and other botanists who regularly exchanged specimens.

Neill's garden was distinguished by an unexpected zoological collection which never ceased to evoke surprise and comment by visitors. He kept a little menagerie of animals to which he was deeply attached. The composition of his animal collection varied over the years but between 1823 and 1835 it comprised a skua, kittiwake, lesser black backed gulls, a cormorant, a gannet or 'soland goose' as it was called, a great black backed gull, (of which more anon), one or two herons, doves, ducks, bantams, a snowy owl blown ashore in Orkney in a snowstorm, an eagle from the Western Isles kept in a large cage, an uncommon type of parrot, an ichneumon, the siren from the marshes of South Carolina (already noted), a toad which lived in a frame and kept the slugs down, a tortoise, a hedgehog and of course cats and dogs, all living in harmony. Loudon also mentioned a southern European frog kept in a large jar provided with water at the foot and a ladder on which it preferred to sit, unless the weather threatened rain when it would descend into the water, splashing about and emitting a loud cry.[13] Most of the birds had been brought to Neill when young and soon became tame, living in the garden for many years. The great black backed gull was an unlikely candidate for such a domestic life. Audubon, in his *American Ornithological Biography* quoted a letter from Neill, summarised as follows.[14] In the summer of 1818 a big 'scorie' (i.e. a young gull) was brought to him by a Newhaven fisher boy, who had caught it in the Forth. It soon learned to feed on potatoes and kitchen scraps and in the company of some ducks became more familiar than them, peeping in at the kitchen in the hopes of a bit of fat meat which it relished. It used to follow Neill's servant Peggy Oliver, calling to be fed. After two months, the dark plumage revealed it as a handsome great black back. The couple of lesser black backs in the garden never let the newcomer associate with them. Since it was so tame Neill did not cut the quills of one wing to prevent flight. In the winter of 1821 it got a companion in the shape of a heron, wounded at Coldingham and kept at the Old College until given to Neill by the janitor. The heron also became quite tame, with the run of the whole garden and access to the loch. It was still there in 1835. In the spring of 1832 the big gull went missing and it was assumed it had flown off and unlikely to be seen again. But one day in October of the same year Neill heard his servant calling, 'Sir, the big gull is come back'. Neill found it walking about its former haunts in the company of its old friend the heron. After the temporary absence it became more cautious but still picked up herrings and other food left for it. In the following March it disappeared but returned in the autumn. Such winter visits to Canonmills and summer excursions continued for many years. In the spring of one year it returned, accompanied by a young gull – probably its offspring – which was soon shot by local sportsmen who frequented Canonmills Loch in the hopes of a pot shot. The big gull became a celebrity, known to local boys as 'Neill's Gull' and more than once owed its survival to their protection from passing sportsmen.

Public Gardens

Neill was the natural choice for advice on how the town council should proceed in making arrangements for garden sites in the city for which it was responsible. By far the most important of these at the time was East Princes Street Garden. After the draining of the Nor' Loch, this piece of ground in the middle of the city became something of an eyesore, to which the council turned its attention in the 1820s. The site had become a place where local housewives bleached their linen – a practice which appalled some citizens – while others upheld it as a legitimate public right. Others had their eye on it as a nursery or even as a cricket ground, while there was always a threat of building. Faced with such conflicting alternatives the council set up a committee with Neill as convener, to make recommendations as to how the site could be embellished.

On September 2 1829 Neill presented his committee's recommendations.[15] The site would be greatly improved by the planting of trees and shrubs, with exclusion of the public until the trees had grown big enough. Next to the iron railing along Princes Street there should be a 20 foot border, trenched 18 inches deep and provided with good soil. This could be planted with a row of lime trees at 10 feet apart, with a subsidiary row of English elms and sycamores, the gaps filled with holly, laurel and privet. Immediately adjacent to the border there should be a 20 foot wide gravel path, for pedestrians only. The plan envisaged an avenue of tall trees which allowed views of the Old Town from Princes Street. South of this walk, and at the top of the sloping bank, there should be a ten foot wide border, also with a principal row of limes, together with elms, sycamores and hardy evergreens. The general aspect would be improved if a roughly triangular space at the south apex of the site were also planted with hardy trees. The report listed the different sorts of tree they had in mind. The limes, sycamores and elms would need to be of the largest size kept in nurseries, i.e. four year transplanted. They did not anticipate their cost to exceed £7. The biggest outlay would be the cost of palings to protect the new planting. The committee thought the whole job could be done for not more than £70. The council accepted these recommendations.

On April 12 1831 the committee was pleased to report that the undertaking was almost completed at comparatively small expense, since the planting and erection of paling was carried out by the town's labourers and carpenters.[16] More than five acres had been planted with nearly 27,000 trees and shrubs. Such dense planting was needed since the soil was of such variable quality that not all would survive. Many of the large, quick growing sorts like willows and poplars were included as nurse trees. The site was mid-way between the Old and New Town and exposed to smoke which was not tolerated by spruces, pines and Portuguese laurel. The committee could report that an area 16 feet wide by 820 feet long had been planted with a double row of limes and elms, as proposed. A five foot wide path had been laid by long sweeps along the different banks to provide access to the plants. The central meadow (the site of the former loch) had been limed and planted with grass, while the 'offensive place of retreat' – i.e. the public privy in the north west corner – had been demolished. To protect against continuing 'nuisance' the site had been planted with wild brambles, briars, blackthorn and seeded with whin.

A notable feature of the enterprise was the free gift of trees and shrubs from many different sources. Neill first applied successfully to likely potential donors. The West Princes Street Proprietors donated 300 sycamores; the Professor of Botany, Professor Graham, produced 1500 plants from the Botanic Garden while the Experimental Garden of the Caledonian Horticultural Society produced 500 plants, including 250 bay laurels and 50 lilacs. But more was needed so Neill approached local nurserymen and was overwhelmed by their response. In the words of the report, Messrs Dickson of Waterloo Place led the way with 1200 specimens – mostly forest trees which included willow and poplar and also 5000 spruce, fir and Scots pine. After enquiring what else was needed they sent along more than 2,500 ornamental trees which included turkey oak, striped gold leaved plane, fern-leaved chestnut, purple beech, double scarlet thorn, guelder rose, Siberian lilac, privet and balm of gilead firs. The lime avenue was supplied by James Dickson & Co. of Inverleith Nursery, which also provided 1000 trees including turkey oak, mountain ash, walnut, hazel and bird cherry. Eagle and Henderson provided 200 Portuguese laurels, trained in the tree style. Three other nurserymen – Thomas Cleghorn, Alexander Wight and Charles Lawson – also made generous donations, together with a private benefactor – John Bonar of Kinnerghame Berwickshire – who provided 8,500 forest trees. As a result of this deluge of gifts, the cost to the Council was less than £5. Such a widespread and generous response reflected the high standing Neill enjoyed in the local horticultural community. The lay-out and planting – under his guidance – had succeeded in embellishing the site and solving the council's problem of how to manage it. A view of 1843 from the Mound shows neatly laid out paths, flower borders and well established trees. But, sadly, it was not to last much longer. The Edinburgh and Glasgow Railway wished to extend its line from Haymarket to the east end of Princes Street, for which permission was granted in 1844. Construction of track and tunnels swallowed up about a quarter of the garden site, including the dense tree planting, while much of the rest was occupied by workmen during the construction of the extension which opened in 1849. The original scheme was further wrecked by the building of the Scott Monument – inaugurated in 1846 – so by the end of the decade, there was little left of what Neill's committee had effected. It must have been a source of acute disappointment for Neill to witness in the last years of his life, the destruction of what he had laboured so hard to achieve. Of course later, the East Princes Street Garden was redesigned to great effect but he was not there to see it.

There is also evidence that Neill – in association with Professor Graham – provided advice in the design of the Carlton Hill pleasure ground. Neill's name also crops up in the negotiations relating to West Queen Street Garden but there is insufficient detail to determine the degree of Neill's contribution, although given his roles as secretary of both the Horticultural and the Wernerian Society, it might be expected that his views were often sought informally on such civic developments.[17]

Articles on horticulture

Neill produced at intervals comprehensive, practical guides to gardening and orchard management. Naturally they had much in common. The first of these appeared in Brewster's *Edinburgh Encyclopaedia* (1830)[18] and the second in the 7th edition of *Encyclopaedia Britannica* (1842).[19] This was substantially reproduced and extended in his *Fruit, Flower and Kitchen Garden* (1845)[20] which went through several editions and was both widely appreciated and plagiarised. It provided a complete practical guide to horticulture in all its forms. He included a detailed guide to fruit culture, with advice as to soil, exposure, types of wall – including hot walls – the management of fruit trees, pruning, grafting, different forms of training of the branches followed by a general survey of the different fruits and their principal varieties. The culture and varieties of small fruit such as currants, gooseberries and strawberries were considered in similar detail. A considerable section was devoted to the forcing garden, with a discussion of alternative heating arrangements, the alignment of flues, glazing, the design of the peach house and vinery, the culture of melons and cucumbers and the uses of hot beds together with an extensive treatment of how to achieve success with pineapples. The kitchen garden was systematically described with sections devoted to the cabbage tribe, legumes, root crops, onions, salads, herbs and every other vegetable that anyone would want to grow, together with the most suitable varieties of each. The flower garden was treated just as thoroughly with advice as to soil and the lay-out of beds and paths, followed by a guide to the species and varieties of shrubs, perennials, annuals and bulbs, grouped according to season of flowering. The text was rounded off with a calendar of what should be done in successive months of the year. It was a very comprehensive and practical guide which provided inspiration and example for subsequent writers on gardening.

The Railway Battle

In the late 1830s Neill was faced with a potentially disastrous threat to his garden. The Edinburgh, Leith and Newhaven Railway Company was planning an extension which would drive a tunnel under his property. He immediately launched a vigorous attack on the plan, hoping to enlist the support of other property owners likely to be affected by the scheme. His first move was to produce 'A Pamphlet addressed to the Commissioners of Improvement, the inhabitants of the eastern division of Princes Street and of those streets under which the tunnel is proposed to be carried and to the Shareholders of the Railway Company'.[21] It went through three editions, sold at sixpence and was widely circulated. The third edition was dated 1837.

In the spring of 1836 a bill had been introduced to Parliament for construction of a railway from Edinburgh to Leith and thence to Trinity and Newhaven. Neill asserted there were great problems which had so far attracted little notice. The town council had opted for a neutral stance and did not wish to become involved. As a member of the committee of superintendence of George Heriot's Hospital, he expressed concern at the damage to the Hospital's valuable fields by deep railway cuttings, but was assured the banks would be 'ornamented' by shrubs. In the railway company's document of November 17 1835 the route was described as running from the foot of Scotland Street to Heriot's Hill, right through his property. Since it was contrary to parliamentary practice to sanction access to private gardens he could not understand why his little 'botanic garden' was to be cut in two, his hot-house destroyed, some of his best trees uprooted and his dwelling house undermined by the proposed tunnel.

The bill relating to the railway was read a first and second time and then sent to committee to which he decided to appeal for protection. Following advice, he went to London to advance his cause. He was kindly received by two members of the committee, the Honourable Members for Leith and Newhaven, but the promoters of the railway were hostile. A petition from the proprietors of Scotland Street arrived too late for presentation. The landlords of East Princes Street remained silent. In spite of considerable opposition, Neill's petition was presented.

The committee recommended firstly that the railway company was prohibited from entering or occupying any part of his premises situated to the west of a line designated on a specific plan, endorsed by the chairman of the committee. Secondly, where the tunnel passed through his ground, it should be constructed without disturbance of the surface – the operation to be supervised by an impartial and competent civil engineer. Thirdly, Neill was awarded the option up to six months after the opening of the railway, to compel the company to purchase the whole of his property at Canonmills if he found he could no longer live there. The railway company strongly opposed awarding him the privilege of a jury valuation and they won their point in the committee, but when the proposals went to the House of Lords, Lord Shaftesbury saw no reason to refuse a jury valuation and with this amendment the bill passed and received royal assent.

As soon as Neill learnt of the ruling he began to make plans to vacate the space east of the designated line. This would entail removal of a stable, a poultry yard and also his eagle house, as well as the demolition of two old tenements belonging to him, adjacent to the garden. He was not told when operations might begin. Several months later he was informed that the company had decided to adopt a new route for their line, further to the west, and this would not touch his property. They intended to send the line on arches across Canonmills Loch. When Neill protested that under the terms of the Act they could not make such drastic changes, he got the reply 'Nous verrons'. Concerned at the apparent impending threat to the loch, he observed that he and his predecessors had long enjoyed the privilege of using water from the loch. He also noted that the plans conflicted with an Act of 1827 against the erection of buildings of any kind upon the area of Princes Street east of the Mound.

The railway company then issued a pamphlet in which they rejected and ridiculed all his arguments.[22] They asserted that what really bothered him was his loss of the chance of making a handsome profit by compelling the company to purchase his property and that all his expressions of concern for the common weal were a sham. They were uncompromisingly rude. Neill had talked about the 'amenity and privacy' of his little villa, no doubt brightened in his imagination by what he hoped to extract from the company. What were its charms, 'situated among the mephitic exhalations and noisome odours of Canonills Loch, a common nuisance and drowning place for cats and dogs?' In passing they even denied his right to take water from the loch.

In replying to this diatribe Neill issued a further pamphlet in 1839, utterly rejecting his wish to sell his property at a profit.[23] He instanced previous correspondence in 1837 with the secretary of the railway company who had repeatedly tried to induce him to sell – even offering to have the sale price agreed by jury and hence likely to result in a high price – but having collected so many rare plants and spent so much money on the garden, no price offered by the company would induce him to sell. He thought it very paltry of the directors to underestimate the value of the property, while at the same time, urging him to sell. 'May not their mouth-piece be likened ... to some old clothesman magnifying the defects of articles he wants to buy, and when refused, resorts to coarse invective?' He went on to accuse the directors of misleading Parliament for they had altered the level and line of the proposed railway at their own convenience. Their sole object was private gain. The directors had sanctioned a pamphlet full of groundless statements and pitiful evasion, from which 'as individuals they would shrink with disdain'. He ended: 'I now bid adieu to the directors and leave the unscrupulous and coarse-minded author, whoever he may be, to his "glory". This parting shot in the war of words was sent to members of Parliament and widely distributed.

The anguish Neill experienced at the risk of damage to his garden and destruction of the pleasant amenity he enjoyed at Canonmills can hardly be exaggerated. Although he was not entirely satisfied with the outcome, his attack and defence by pamphlet and his appeal to Parliament preserved the integrity of both his garden and premises. His case won support from friends and sympathisers. It was no coincidence that Loudon's and Don's public endorsements of the unique quality and value of his plant collection appeared when they did. They were part of the battle to save Neill's 'mini-botanic garden'.

1   Patrick Neill, *Horticulture, Encyclopaedia Britannica*, 7th edition, vol. XI, 1842, pp.630–691.

2   ----------, *Horticulture, The Edinburgh Encyclopaedia, Conducted by David Brewster with the Assistance of Gentlemen Eminent in Science and Literature*, MDCCCXXX, Vol. XI, pp.177–313.

3   ----------, *Considerations Regarding the Edinburgh, Leith and Newhaven Railway*, Garden Inventory by Neill's gardener, William Lawson. Ry, IV, e.9, National Library of Scotland.

4   J. C. Loudon, *A Note of the Garden of Cannonmills Cottage, the Residence of Patrick Neill, Esq., LL.D., F. L. S. etc. with Lists of the Rare Plants Cultivated or Figured and Described from it. Drawn up from Cultivation Records from Professor Don and Mr. C. H. Smith and Others, 1836, Gardeners Magazine*, Vol. 12, pp.329–341.

5   William Jackson Hooker, Director's Correspondence with Patrick Neill, Kew Gardens Library, London.

6   Walter Calverly Trevelyan, Correspondence with Patrick Neill, University of Newcastle-Upon-Tyne Library, Newcastle-Upon-Tyne.

7   I.C. Hedge and J.M. Lamond, (eds), *Index of Collectors in the Edinburgh Herbarium*, 1970, p.113, H.M.S.O., Edinburgh.

8   E. Charles Nelson, Dr Patrick Neill's Herbarium of Scottish Garden Plants in the National Botanic Gardens, Glasnevin, Dublin (DBN), 19??, Bot. J. Scotl, 46 (2), pp.347–380.

9   Neill Letters, The Linnean Society, London.

10   *Ibid.*

11   *Ibid.*

12   *Ibid.*

13   J. C. Loudon, *Gardeners Magazine*, 1836.

14   John James Audubon, *Ornithological Biography*, Vol.3, pp.312–315, 1835, Adam & Charles Black, Edinburgh.

15   Minutes of Edinburgh Town Council, September 2 1829.

16   *Ibid.*, April 12 1831.

17   Connie Byrom, *The Edinburgh New Town Gardens*, 2005, pp.20, 216, 332, 325, Birlinn Limited, Edinburgh.

18   Patrick Neill, *The Edinburgh Encyclopaedia*, 1830.

19   ----------, *Encyclopaedia Britannica*, 1842.

20   ----------, *The Fruit, Flower and Kitchen Garden*, 5th Edition, 1845.

21   Patrick Neill, *A Pamphlet Addressed to the Commissioners of Improvement, the Inhabitants of the Eastern Division of Princes Street, and of Those Streets Under Which the Tunnel is Proposed to be Carried and to the Shareholders of the Railway Company*, 3rd edition, 1837, Edinburgh, National Library of Scotland.

22   *Reply from the Directors of The Edinburgh, Leith and Newhaven Railway Company to Certain "Remarks on the Progress and Prospects of the Company by Dr Neill, Cannonmills"*, 1839, National Library of Scotland.

23   Patrick Neill, *An Examination of the Reply from the Directors of the Edinburgh, Leith and Newhaven Railway Company*, 1839, National Library of Scotland.

8 🌿 *A Continental Foray*

At the annual general meeting of the Caledonian Horticultural Society in 1815, Sir John Sinclair proposed that the society should commission two representatives to visit Northern Europe to discover what advances had been made in horticulture during the long period when British visitors had been unable to visit that region. With the Battle of Waterloo and the end of the French war, the way was now open to visit the continent again and re-establish contacts. His suggestion was approved and the two secretaries Walter Nichol and Patrick Neill were chosen to carry out the task. Before the trip could take place, Nichol sadly died and in due course it was agreed that Neill should be accompanied by James Macdonald – who was the Duke of Buccleuch's head gardener at Dalkeith – and John Hay – who had formerly worked at Archerfield and was now the leading garden designer in Edinburgh. They were advised to visit Flanders, the Netherlands and Northern France and report back to the society. One of the principal aims was to identify varieties of fruit trees, vegetables and ornamental plants which would find a place in the projected Experimental Garden. They were also to establish contact with growers and nurserymen, with an eye to future importations of new varieties. They completed their mission over a period of nearly three months in the late summer of 1817. Neill had intended to produce merely an internal report but was persuaded to describe their visit in a book for a wider audience. Due to illness and other factors – including a separate visit to France by Neill in 1821 – the book did not appear until 1823.[1] Neill's account is of considerable historical interest. It provides a picture of the state of horticulture in northern Europe, with particular attention to the varieties of fruit of all kinds which were then widely grown. The visitors' comments on what they found novel in the foreign scene often reveal aspects of Scottish life and attitudes by way of contrast.

On August 1 1817 they sailed from Leith in a commodious vessel called *The Czar,* which carried them to the Thames Estuary in a little over four days. They had a few visits to make in London; to Covent Garden and the 'forcing-premises'of Isaac Andrews on the recommendation of the president of the society, the Earl of Wemys. Here they were astonished by the extent of the hot-houses, vineries and hot-bed frames, geared to cater for metropolitan luxury. Andrews specialised in black hamburgh and white sweet water grapes and had large quantities of the Virginia strawberry, growing in about 7000 pots. This was the same variety as that sold in Edinburgh. The visitors' only caveat was at the general untidiness of the premises.

On August 7th they visited the garden at Lambeth Palace where they found the normally abundant harvest of fruit to be very meagre that year. The same was true in Scotland. Macdonald recounted how at Dalkeith, the spring blossom had failed to set and blossom was still sporadically appearing as late as the beginning of August. This was attributed to the cold and wet of the preceding year. At the Mile End nurseries – where John Hay had spent some time 35 years previously – they admired the famous gingko tree, now 50 years old.

The following day they set off for Dover via Canterbury. Travelling through the countryside they approved of the method of harvesting corn with a scythe rather than the sickle used at home. They were astonished to see a team of four strong horses drawing a plough, with two ploughmen in attendance. In Scotland the same task would be accomplished with just two horses and one ploughman. Similarly they saw three horses yoked to a drill, whereas at home one horse would have served. They attributed such extravagance to habit. Neill – ever the sharp-eyed field botanist – kept an eye out for wild species of plants which were scarce or unknown in Scotland. He admired the single red carnations which decked the walls of Rochester Castle. At Canterbury, in the vicinity of the Abbey of St Augustine, they saw the remains of the old monastic garden: ancient vines, an old black mulberry tree and several walnuts. At Dover, while waiting to board ship, they botanised along the foot of the cliffs and marvelled at the quantity of wild cabbage, which was gathered as a pot herb by the local people.

Setting sail in a post-office packet in the evening, they arrived at Ostend the next morning. After customs they put up at the Rose Inn, run by an Englishman. This being Sunday, they experienced their first contrast with the Scottish Sabbath for many shops were open and, soon after midday the sound of fiddles and dancing convinced them they were now indeed in a foreign country. They noticed that the public notices were in both Flemish and French and occasionally – especially above inns catering for sailors – curious versions of English, such as 'Spiritual Liquors' or 'All sorts of drinking sold'. They were not overly impressed by the sparse use of available ground for potatoes or other vegetables.

From Ostend they proceeded to Bruges by horse-drawn barge on a wide and well maintained canal. They put up at a very good inn, to which Louis XVIII had retreated after abandoning Lisle in 1815. They found Bruges a comparatively deserted city, a sad contrast with the days when it was once the emporium of northern Europe, ranking even above London. They visited the best garden in the district – belonging to a wealthy merchant – and found it laid out in the Flemish style, with regular walks, barricades of pleached lime trees with periodical window-like openings, and a fine lake that was – in Neill's opinion – marred by the practical jokes. If the visitor sat on a seat it was ten to one it would sink under him or, if he entered the grotto a concealed pipe would soak him with water. In the fruit garden they saw pear and apple trees trained *en pyramide* and *en quenouille,* planted at eight foot intervals.

They found succory grown on a large scale in every kitchen garden and sometimes there were whole fields of it. This was a novelty for the visitors. The plants were raised from seed every spring. The early tender leaves were used as a salad, while the full-grown leaves were cut once or twice for the benefit of the cows. When the roots were the size of small carrots they were scraped and then boiled and eaten with potatoes and a butter and vinegar sauce. During the Napoleonic wars, the roots were dried and ground up to make a substitute coffee. The visitors noted that both the common winter wheat and the red wheat were grown while the old Roman spelt was also grown in a few places and valued for making pastry.

This region was the home of the broad clover (*Trifolium pratense*), locally known as marsh clover from its preference for moist sites. It was manured with ashes brought from Rotterdam. A great deal of seed was produced for export. In Scotland clover seed was also produced in favourable seasons, although it was hard to separate the seed from the husk – a difficulty that Neill thought could be overcome mechanically. In places like the Carse of Gowrie farmers could set aside for seed an eighth or tenth of their clover crop with the aid of a movable fence. In 1819 very good quality seed had been produced. A farmer in North Berwick had employed women and children to gather the ripe heads by hand. One person could harvest eight pounds of seed in a day.

Of particular interest was the garden of the Capucin friars, who had managed to retain their monastic garden during the Revolution. Perhaps their contempt for money saved them. The very old garden was in the charge of a venerable gardener: poorly dressed, with a large hooked, pruning knife attached to his button-hole, wearing a greasy woollen cap and a dirty white apron, unlike Scots gardeners who always wore a blue apron. Among the several varieties of pear, they found the Orange Bergamot and varieties which the gardener called Casserine and Callebush as well as Longue Queue of Louvain. The Passe-colmar – which was believed to have arisen in this part of Flanders – was particularly abundant. The fruit was late but of large size and fit for eating from December to January. A sort of pear which the gardener called Cheneau looked rather like a Gansell's Bergamot.

From Bruges they went next to Ghent, again by canal on a barge which was fitted up with a deck awning. The numerous English travellers sat on deck while the Flemings kept to themselves in the cabin. They were served an excellent lunch aboard. Ghent was the birth-place of the Emperor Charles V and was then the capital of Austrian Flanders. At the princi-pal gate of entry they had to show their passports, but at the mention of the name Macdon-ald the official appeared 'electrified', for he hastily scanned through their papers and left his desk to inform them that his ancestors were Scots. The travellers were glad to find that the weather had improved and it was now quite warm, confirmed by the reading on Mr Adie's pocket thermometer which had been presented to Neill by its maker before they left home.

Their first objective was the vegetable and fruit market where they saw a vast quantity of haricots and a beautiful red kidney potato, samples of which were purchased to take home. They were interested in a plum, shaped like a cherry, known as the Mirabelle – widely planted in hedgerows and used by the Flanders nurserymen as a stock for peaches and nectarines. Their next port of call was the Botanic Garden where the gardener advised them to visit the de Cock nurseries.

The following day they found all the shops shut, processions in the streets and people streaming to church, for it was the *Fête d'Assomption*. They joined them in the Cathedral of St Bavo. The Catholic service was new to them, especially the priests with their rich vestments and the altar boys swinging censors of incense. Part way through the service Neill and Macdonald discovered that Hay had vanished. The display of popish pageantry had been too much for his Presbyterian sensibilities and he had fled.

At Smedt's market garden a great deal of effort was devoted to the production of seeds for sale. However the different sorts of lettuce, carrot and onion were grown is such close proximity that hybridisation was unavoidable and no doubt contributed to the variation commonly encountered among samples of seed from this district. Their next objective was the orchard of the suppressed Abbey of Bandeloo. Napoleon had decreed this was to be a centre for botanical education. The garden occupied more than three acres and, although quite well stocked, hardly competed with the Edinburgh Botanical Garden in quantity and variety of species. At the commercial nurseries of de Cock they found a variety of fruit trees. The apricots included Le Grand Maréchal and La Crapaude. There was nothing remarkable about the plums, which were chiefly represented by the Longue Bleau – locally popular but inferior to the Blue Magnum. There were many pears which included several types of bergamot, Crasanne and Grande Bretagne (a fairly large winter pear), Colmar d' Hiver, Passe-colmar, Poire Capucin (for stewing only), Angelique de Bordeaux, Poire de Couvain, Beau Present (an excellent and beautiful pear) and Mansuette Gris, which appeared to be close to what was known at home as the 'grey achan'. Others included the Duchesse d'Ardenpont (a large winter pear) and Beurré d'Ardenpont (which was a new variety). Ardenpont was a village near Tournay, noted for its hardy fruit. The apples were mainly Calvilles, Rennets and Carpandies (Court pendu).

While at Ghent they visited an exhibition of paintings by contemporary Flemish artists. This was crowded with local people in their best clothes which were too garish for Neill's taste for black and sombre hues, since their upper and lower garments often differed wildly in colour. But the people were wonderfully well behaved and decorous, prompting him to reflect that this would hardly be the case in Scotland where it would be hazardous to allow the general public access to valuable works of art – although he thought it would be a good idea to have an exhibition of painting by living artists in Edinburgh.

Armed with a letter of introduction from Sir John Sinclair, they visited the Agricultural and Botanical Society, which had been founded in 1809 and annually ran two Festivals of Flora: one about mid-summer and the other in mid-winter. They had a pleasing custom whereby the flowers themselves were the competitors and were awarded the prizes.

Hiring a carriage they made for the garden of the principal banker along a road lined with forest trees, invariably straight and regular. Here they thought the gardener presented a woeful sight, with bare legs and huge wooden clogs which a Scottish gardener would find a great encumbrance. There were many walnut trees in fruit and several fine specimens of the variegated elm, up to 30 feet in height and grafted on ordinary elm. A serpentine walk lead to an artificial cave paved with the meta-tarsal bones of sheep. They noticed a dibble with nine equally spaced holes for planting peas or beans. The greenhouse was untidy and full of lumber – a state of affairs they often found. About midday they travelled to Leerne (about ten miles from Ghent) along tree-lined roads where the young trees were often protected from passing cattle by twining bramble round the stems. They remarked on the small fields divided up into little plots with different crops – a result of the laws of inheritance that led to such sub-division of land among the heritors. They were surprised to see broom grown as a crop. The young flowers were gathered in spring and pickled as a substitute for capers.

They next visited a fine old Flemish chateau where the grounds were laid out in the Dutch style and had been modelled to make banks that sloped to a lake. The garden was tastefully laid-out and supplied with greenhouse, orangery, peach houses etc. There were a variety of pears, but Neill thought the soil here was too light and better grown specimens could be found on walls in Scotland. But the chief ornament was an immense meadow spreading before the house and grazed by hundreds of fine cattle belonging to the people of the nearby villages. Sometimes the cattle had a piece of basket work over the muzzle, which did not hinder feeding but prevented damage to trees.

Repairing to the village for lunch they remarked on the coarse and clumsy furniture, the snow-white napkins and large silver forks, in contrast to the knives which were no better than 'threttypenny joctelegs'.

The next day being Sunday, they went to a Protestant church which had been established for the benefit of the Dutch troops, having been taken over from the Capucins. Sentries were posted both outside and within the church, to guard against the excessive zeal of the Flemish friars. The pews were supplied with large quarto Bibles, inscribed with a notice that they had been presented by King William.

Later in the day they visited the garden attached to a boarding school for young gentlemen, who included a number of boys from English Roman Catholic families. As a hobby, the headmaster had developed an extensive nursery where he had 30,000 young fruit trees. They ordered a dozen or so pear and apple trees for the Experimental Garden. The owner remarked that out of a great many seedling peaches he had raised, only one was of superior quality and promised to let Macdonald have one. When they returned to their hotel, they found the adjoining square occupied by rope-dancers, tumblers, 'merry-andrews' and a throng of people – a far cry from the sobriety of a Scottish Sabbath. evening. This place was infested with beggars who would intone 'Ah myn heer' or 'Pour l'amour de Dieu' as they passed by.

The following day they visited several gardens, the first of which boasted a huge stand of *Lilium superbum* and a good collection of evergreen American shrubs, so highly esteemed by British gardeners. Yet again they found the gardener wanting for he did not know the names of many of the plants. The focal point of this garden was a rotunda with six Ionic columns on an eminence with a fine view. Below the building was a cave with a sculpted female figure, reclining in an attitude of grief.

The next garden belonged to a Madame Vilain Quatorze, where for a change, they were favourably impressed by the lively manner and appearance of the gardener, as well as the extensive greenhouses. There was no tax on glass and this accounted for the wealth of glazing. The house was surrounded by a moat and connected by a splendidly decorated gallery to the largest greenhouse, whose interior was revealed by a large mirror in the house and lesser mirrors distributed among the plants. In winter lamps were hung in the gallery and the centre of the greenhouse to charming effect. The only discordant note was the presence among the plants of huge basket-work figures clothed in the full dress of the King of the Netherlands. Among the excellent collection of West Indian plants, Neill noted the papaw and true to didactic habit, described the properties of the juice for tenderizing meat so that beef and mutton could pass for veal or lamb and old poultry made palatable as chicken. In a farm outside the chateau they were glad to see a large field of hemp, seldom or never seen in Scotland, as well as a field of Indian corn, which must have been near its northern limit of cultivation.

During this outing they were plagued by two unsavoury characters that dogged their footsteps, boasted of having served in Napoleon's army and doing their best to irritate them. Suspecting they might be attacked, the visitors held themselves in readiness. They thought it likely that these unwelcome companions may have been government spies.

Neill and his friends were very favourably impressed by Ghent and wished they had more time there. They greatly approved of the cleanliness and order of the Ghent meat-market – unlike that of Edinburgh, where the gutters often ran with the blood of newly killed pigs and the sounds of turkeys and ducks being slaughtered assailed the ear. They learned that Sir John Sinclair had been recently responsible for introducing rutabaga to Ghent. One of the local customs that caught their attention was that of placing small mirrors at the angle of windows, so that anyone sitting inside near the window had a view of what was happening in the street outside. Archery was a favourite pastime. An imitation bird was fixed to the end of a long pole and this popinjay became the target. Neill and his companions were surprised to see horses confined in a strong wooden frame as if they were to have a tooth drawn or to be bled, when only the fitting of a shoe was in hand. They remarked that the manufacture of lace had been quite superseded by the making of cotton goods – of no benefit to the morals of young women who were brought together in large numbers, whereas formerly they worked on lace in their parents' homes. Newspapers were conspicuously scarce.

Antwerp was next on their itinerary. They left Ghent by coach at five in the morning. After travelling for three or four hours they enquired about a stop for breakfast but this request fell on deaf ears and they were told it was not the custom to do so. When they reached Antwerp, after finding an auberge, they did a little sightseeing. They found the very fine cathedral somewhat marred by the array of cheap stalls ranged outside, reminding them of what had formerly been the custom at St Giles in Edinburgh, before the stalls were banished. They were agreeably surprised by the well-grown trees, which added greatly to the amenity of the town. They visited a garden which was laid out in a mixture of Dutch and English styles, but rather bizarrely embellished in different areas by a flock of life-like stone sheep, a sculpture of a bull being attacked and elsewhere, grotesque human figures including 'a drunken fellow in a situation which could only evoke disgust'. There was also a fanciful tomb and adjoining cave – supplied with the usual pipery to soak the unwary, but fortunately (as was generally the case) the stop-cocks and taps were out of order. There was even a chair designed to soak the nether regions of anyone who sat on it.

In the fruit-garden they came across a number of Beurré-type pears little known in Britain, such as Beurré d'Or, Beurré Royale, Beurré d'Hiver (possibly the same as

Chaumontel), Beurré Blanc and Bezie Bleu. The pear generally known as Brown Beurré in Britain, was here known as Beurré d'Angleterre. Virgouleuse (an ice pear) and Bergamotte de Paques were favourites in Antwerp. Though uncommon in Scotland, these were well known in England. Délices d'Ardenpont and Passe-colmar were excellent. Belle de Bruxelles was also a recent, promising introduction. Nouvelle Epine d'Hiver, a seedling raised at Antwerp, was another. The visitors hoped to obtain grafts of a number of these for the Experimental Garden. There were several apples at Antwerp unknown or little so in Britain, such as St Jean d'Or, Red and Grey Rennet, Mountain Rennet, Fine Verte, Calvin Acote, Pomme Royale, and Drap d'Or (which was neither Fenouillet Jaune nor Golden Rennet – both of which were sometimes known by this name). There were different varieties of Court Pendu such as Court Pendu Peche, Court Pendu Rozart and Fossette – all large dual-purpose, well-flavoured apples which would keep until May or June. There were no novelties among the plums. The best was the Reine Claude or Green Gage, while the commonest was the Hungarian Blue egg plum.

In the course of walking about Antwerp, they commented on the number of effigies of the Madonna and Child in the streets. These had been swept away during the Napoleonic era but had been restored by the King of the Netherlands, even though he was a Protestant. When they attended a service at the cathedral, they were surprised to discover that there was no separation of classes. All grades of society mingled at the service. The ragged porter kneeled beside the well dressed merchant. This was another departure from what they were used to, where many 'a bustling beadle is employed to prevent the general part of the congregation from being intruded on by worshippers of humbler degree'.

They again remarked on the dearth of newspapers. There was only one small folio sheet which expressed the greatest antipathy to Britain. Apparently the defeat of Napoleon had been greatly regretted by the Brabantines. Copies of a paper which printed a protest by the ex-Empress were suppressed and a military guard posted at the stationer's premises where the newspaper had appeared.

Leaving Antwerp, en route for Rotterdam, they went via Williamstadt by dreadfully muddy and rutted roads, prompting Neill to lay down two guiding rules for travellers to the Netherlands. Firstly, never travel after heavy rain for the roads were almost impassable; and secondly, take a supply of provisions since the coach service entirely disregards the comfort of passengers in that respect, quite unlike the English coach service where the traveller was taken care of in every way.

At Rotterdam they found that the autumn fair or 'karmes' had just closed. This reminded them of the Cranes of Edinburgh, which probably arose as an extension of the All Hallows Fair and had recently been discontinued. Many Scottish merchants had settled at Rotterdam and they found themselves received 'as a countryman would expect to be received'. Punctilious as ever, they went to the Protestant Church on Sunday, where they found the service much the same as at home but with a few antique touches. Thus, the sandglass – formerly a feature of the Scottish pulpit – was still in use here. Also, while waiting for the minister to appear, the clerk would read aloud from the Scriptures, another custom whose passing in Scotland was regretted by Neill. After the service, there was a parade of the Regiment of Burgher Guards, wearing a uniform not unlike that of the Edinburgh Royal Volunteers. They met up with a couple of Scots pastors of the Scottish Church (which was paid for by the Dutch Government).

They were impressed by the size of the barges at Rotterdam. They inspected one from Cologne. It accommodated a shop full of china and crockery, an elegant cabin with several bedrooms leading off, an excellent kitchen and on deck there was a mini-garden of potted shrubs and plants. The family would live on board for months at a time.

Once again they found themselves admiring the fine trees which graced Rotterdam.

Impressed by this sight, Neill wished there was more attention to tree planting in Edinburgh. He thought the appearance of 'the ponderous and clumsy Earthen Mound' would be greatly improved by the planting of trees on its slopes, while the marshy ground left by the draining of the Nor' Loch could be improved by planting willows and alders.

They discovered that they had just missed a party of their countrymen who had passed through Rotterdam. They included Lord Chief Baron Dundas, Sir William Rae (later Lord Advocate of Scotland), Principal Haldane of St Andrews University and Mr Stevenson, civil engineer. A readable account of their trip appeared in issues of *The Scots Magazine* for 1819 and 1820. About the docks at Rotterdam they remarked on the sledges, drawn by large well-groomed, black horses, with a small barrel at the front from which trickled a small stream of water to reduce the friction with the causeway. As usual they visited the vegetable market where they were surprised by the huge demand for cauliflower which was delivered in loaded barges over several months. Scorzonera was a favourite here. They found the various legumes so old and stringy as to be deemed worthless in Scotland. However, it was the custom here to chop them up very finely to make them palatable. They looked in vain in the bookshops for a book on gardening in the Netherlands and were told nothing of that nature had been published in recent years.

They commented on the long hours worked by the merchants at the Exchange. Soon after two o'clock the piazzas were crowded with merchants, raised from their dinner at one o'clock. It was not the custom to take wine after dinner. They went home for tea between three and four and then went back to the Exchange for several hours' further work. The evening supper was the only time for any relaxation. Neill observed that in the early 18th Century, the Edinburgh merchants closed shop for dinner between one and two o-clock and were regaled by the chimes of St Giles.

From Rotterdam they travelled to The Hague by canal – an agreeable way to view the countryside and note the different plants growing by the water's edge. Lapwings and starlings abounded. They found The Hague a large and elegant city, with handsome tree-lined approach roads. They put up at a splendid hotel in a fine square. Here the most immediate impression was of the abundance of orange cockades displayed in the streets. There were very few of these in Antwerp, a few more in Rotterdam but here they were everywhere. Before the French Revolution, the Dutch had no time for the House of Orange. The Burgomasters even made themselves ridiculous, not only by eradicating marigolds from their gardens but they also banned the sale of carrots and oranges, on account of their 'aristocratical' hue. But their attitude was now very different.

At the near-by fishing village of Scheveling they watched the fishing fleet with their multi-coloured sails, driving their broad-prowed boats right inshore, close enough for the fishermen to jump overboard and carry their catch ashore. They commented with approval on the cleanliness of the village and the tidiness of the fisher wives, who compared favourably with their Scottish counterparts. Since they were well on with their trip, Neill and Hay sent off a report to Dr Duncan senior, vice president of the Caledonian Horticultural Society, to be read at the annual anniversary meeting on Sept 2nd.

They were interested to see how well the gardens were stocked with fruit-trees – including mulberries, which were in almost every garden. While visiting the University, they ran into several young Scots who were studying law. Since Britain had acquired the Cape of Good Hope it was necessary for lawyers to be familiar with both Dutch law and language. Incidentally, they marvelled at the continual doffing of hats in Holland to all ladies, known or unknown, and even to strangers. They restricted their doffing to significant subjects like Clusius' palm in the Botanic Garden or the statue of Coster, whom the Dutch incorrectly regarded as the inventor of printing – a particularly appropriate gesture for Neill, the printer.

They visited several *bloemstries* or commercial flower gardens and the garden of the Palace of Haarlem, where they saw apricots growing under glass and also peaches in a glasshouse divided lengthwise by a wooden partition to allow grape vines to be grown on one side. There were some standard apples loaded with fruit. Autumn Calvilles and a variety called Zydehemd or Jerusalem apple were particularly popular here. Raspberries were forced by growing them on either side of a long trench of tanner's bark to benefit from the extra heat. Neill thought it would be worth trying this in Scotland. At Kreps' nursery they found a considerable collection of roses, but as far they could see none of the varieties raised by Messrs Brown of Perth, nor by Messrs Austin of Glasgow were included, although in Neill's opinion, they excelled in delicacy of form any of the recent introductions in Haarlem or elsewhere in the Low Countries. They were derived from semi-double forms of *Rosa spinosissima*, as well as multi-petalled varieties of *R. alba, centifolia* and *gallica*. They saw a great quantity of dwarf apple trees that went by the name of *Pyrus sempervirens* or Bastard Rennet. They were raised in pots on a large scale and were very popular, bearing leaves and fruit for display at grand dinners. In general dwarf fruit trees were greatly sought after in the Netherlands. At Mr Jean Moonens' garden they were shown around by the owner who proudly indicated plants he had obtained from Lee and Kennedy's nursery at Hammersmith – generally regarded as foremost in the world for herbaceous plants. In a pit of tanner's bark, which was being cleaned out by workmen, they saw hundreds of stag beetles in all stages of development. In the fruit garden he noted a great many varieties of apples, mostly with unfamiliar Dutch names, including the Blanke Aagt and the Wyker Pippen, which were particularly recommended. The visitors ordered a couple of year old dwarf trees and other varieties, as well as plants of Blue Frankendale and White Early Candia grapes These arrived in Edinburgh during the winter of 1817–1818, but since the Experimental Garden was not yet established they were planted out in several private gardens. The Dutch were skilled at the provision of ready-made hedges of large shrubs, while trees of 20 or more feet in height were successfully transplanted. During a walk, Hay spotted what Neill termed a large *Boletus igneus* growing on an oak stump. When dried this fungus was known as 'amadou'. It was imported from Hungary under the name of Boomzwam. Sold at every street stall in the Netherlands, it served as a convenient portable tinder.

At the cathedral they encountered a few different customs. Many of the large congregation – especially the ladies – were seated on reed chairs and the handing of these over head-tops for the benefit of 'dames of distinction' afforded polite exercise for the gentlemen. During prayers women knelt and men stood. Men generally wore their hats in church, except during prayers and praise. This often gave offence to English visitors who according to their faith regarded the church as consecrated ground, whereas both the Dutch and the Scots regarded that as superstition. Neill was particularly observant of differences in religious behaviour, no doubt arising from his early upbringing as an Anti-Burgher. There was less coughing in Dutch than in Scots churches. He thought this could not be accounted for by the comfort of foot-stools and the wearing of hats but was due to the greater fluctuation of temperature in Scotland. After the service Macdonald went botanising.

At the fish-market they were interested to see that sea fish like haddock, cod, young coalfish and various flat fish were kept live in tubs. Half a dozen storks were wandering about the market. A small wooden shed had been put up for their protection at night. Goats and dogs were harnessed to pull carts of merchandise or very sedate looking children. A kind of church-officer, dressed in black with a long crape attached to his cocked hat and carrying a paper, was the official announcer of deaths. In a large piece of open ground they met up with a fair that displayed panoramas from Flanders, wax works from Italy, dwarfs and giants from Germany, rope dancers, an exhibition of marionettes from Paris, and John Bull with an exhibition of wild beasts from Exeter Change. There were booths for sale of broodyerties, waffles and small pancakes eaten with butter and sugar. There was also a panorama of Waterloo in which the Prinz van Oranje played a more conspicuous part than either Wellington or Bonaparte.

There was no coal. Instead, two sorts of peat were used, called high and low. The former consisted of the dry upper layer, which when removed, exposed ground suitable for arable crops. The other sort was a kind of mud-peat, removal of which led to flooding of the soil. There was no fresh water for drinking as it had been brought by barge from Utrecht.

At The Hague, Leyden and Haarlem they found *Fragaria vesca* (the wild wood strawberry) was preferred. It was known as the Boskorper strawberry from the name of the woods at Boskorp where the plants were gathered. By careful culture in pots, they could produce fruit from March to November. Every second year the plants were discarded and a new supply was collected from the wild.

At Utrecht – their next port of call – they were glad to have fresh drinking water and be free of the smell of stagnant water. They visited the Botanic Garden, which had a rich collection of South African plants. They also visited the Jewish Quarter, remarked on the many handsome women, the characteristic appearance of the inhabitants and marvelled at the ability of these people to maintain their identity in spite of being so widely scattered. At the fruit market they found the Gravenstein and Borsdorfer apples especially esteemed. The former was a large yellow and red apple with rich and juicy pulp, although not very tender. If gathered before ripe it would keep plump throughout the winter. It was often dried like the Foppen pear. Trees of this variety from Hamburg had been grown by Duncan Cowan, who occupied the ancient Edinburgh garden which belonged to the Regent in the 16th Century and which was then still in the hands of the Murrays. Neill noted that this Edinburgh garden still harboured pear trees of great age, including Longuevilles, Achans and Jargonelles. The Borsdorfer apple originated in Saxony. It had two varieties – the Common or Autumn: yellowish and of middling size; and the Red or Winter Borsdorfer which kept even longer than the former. The blossom was not much affected by frost so Neill thought it might be a suitable variety for Scotland – correctly so as later experience demonstrated. There was also another German apple called the Frau Appel. This was very much a winter fruit of large size with brown skin and hard but tasty pulp. It would keep even to July.

At Utrecht they encountered the Calebasse pear, which was nearly as good as the Jargonelle. It was an early autumn fruit of good flavour. Unknown in Edinburgh, it was another candidate for introduction. Other varieties new to Neill and his friends were the Fig pear, Poire Madame and the Wyn pear. The apples on sale were the Permain Pippin ('pipling') and Koning's Pippin. There were also different varieties of Calville, classed as dessert apples but not of a quality that would be acceptable at Scottish tables. They were told that no seedling apple had been raised during the preceding 30 years.

From Utrecht they went next to Breda. En route they passed acres of medlars – the first they had seen – as well as the rather bizarre practice of painting lime trees in front of a house with alternate black and white stripes. The country was marshy with plenty of grebes, water hens and coots. Suspended over many pools there were basket work structures like large nests, designed to catch ducks. Although much of the country was inundated the ground must have been hard for they saw cows grazing up to their knees in water. The westward sloping side of the dykes were planted with apple trees which were the largest they had seen. The slow passage of their diligence tried their patience. They had left Utrecht at six in the morning but did not reach Breda until seven in the evening, although the distance was no greater than that between Edinburgh and Glasgow.

In a general comment about Dutch gardens, Neill remarked that in both Amsterdam and Rotterdam he found the white paths constructed with broken sea shells, mostly of cockle shells and of a species of mantra. The collection of these shells was part of the fishermen's trade. Armed with a kind of net-shovel, they caught the shells as they were carried in by the breakers. Their catch was then taken to depots and transported to all parts of the country for garden use.

Although Breda was a fortified town – one of the strongest in the Low Countries – they still found room there for the tree shaded palace gardens, with serpentine walks, open to the public. They visited the best local nursery garden where they discovered that the so-called Breda apricot – long esteemed in Scotland – was not recognised by that name here.

They saw the Yut pear, with large dessert-quality fruit, which kept longer than the Jargonelle. This was another one for the Experimental Garden as soon as it was set up.

From Breda they hired a carriage to make for Antwerp. The pace was very slow, understandably so when they learnt that the same horses had to complete the forty mile journey. From Antwerp they made for Brussels, which they reached on Septermber 7th and found accommodation in the old corn market. In the fruit market they wondered whether the Beau Present pear was really distinct from their own Jargonelle. The Caillat Rosat was of large size, but rather wanting in flavour. The Fondante de Brest was the only other pear worth noting and this turned out to be the same as the variety called Cheneau, which they had seen at Bruges. They were unimpressed by the quality of the grapes on sale. Roasted apples and pears were very popular. They visited the frog market where they saw the 'frog-wives' amputating the legs of edible frogs with large scissors. The sight of the still writhing corpses discarded in the gutter disgusted them. In the Botanic Garden there was little of interest, beyond a small collection of medicinal plants and a fine collection of orange trees which were manured with sheep dung. The *Hibiscus syriacus* blooming in the borders proved the climate was milder than in Edinburgh. The Marvel of Peru was a favourite ornamental plant in the parterres.

At the Palace of Lacken they were at once admitted with the comment 'les Anglais peuvent aller partout'. The King passed in his coach so they doffed their hats and were saluted in return. They found the grounds well laid out, in an exquisite setting. It was so hot they retired to a rustic shelter in the public garden of the village to watch the local young men, stripped to the waist, shooting at the popinjay with bow and arrow.

Being so near to the battlefield of Waterloo, they could not refrain from visiting it. They were given a graphic description of events by a survivor. They were scandalised by the attempts to sell souvenirs to the numerous tourists. At the old Dutch elm tree which Wellington chose as a base, boys climbed up to break off branches which they sold to the tourists. They sought to discourage them by refusing to purchase. They made their way to the shabby old house where Wellington and Blucher met after the French were fleeing. Although hardly meriting the title of 'hotel', they were served wine to drink to the two generals and remember the Scots who fell at Waterloo. They did not forget either to pledge their fellow gardeners, who were probably just then met at Oman's in Edinburgh, since this was the eighth anniversary of the founding of the Caledonian Horticultural Society. On their way from the battlefield they were moved at the fragments of human ribs and skulls, kicked up on the roadway – a sight to which the local people had apparently become quite indifferent.

Neill could not refrain from mentioning the Brussels sprouts that had been grown there for more than 400 years. Only sprouts of an inch or less in diameter were accepted in the market. Although grown in the best gardens in Scotland, they had not yet reached the Edinburgh fruit market.

With the aid of a letter of introduction from the Earl of Wemys, they called on the Duc d'Arenberg, who however was out of town. This nobleman had lost his sight as a young man from a shooting accident. Although some of his estates had been pillaged during the Revolution, he had not been harmed and still retained considerable wealth. In his town garden they encountered an orange tree, said to be nearly 400 years old. They were also interested in a special sort of wheeled hoe operated by two men, one at the front pulling and the one behind steering. This was used for cleaning paths and could do the work of six men.

At another garden, the gardener named without hesitation St Germain, Colmar and Grande Bretagne as the best three pears. They knew the first two of these well enough but the third turned out be the familiar Grey Achan, honoured here with a west wall beside the Chaumontel. At this point, Neill was moved to reflect that while valuing the Calebasse, the Yut or the Passe-colmar, they should not neglect the Muirfowl Egg, the Warden, the Ballencrief, the Pollockshaws and other Scottish pears. One of the first projects of the Experimental Garden would be to compare them for quality and sort out the confusion about their names. Hay drew their attention to the small size of the peach trees, which were almost all young. They were told that peach trees seldom lasted beyond seven or eight years, due in their opinion to bad pruning.

A highlight of their visit was the nursery garden of Professor van Mons – well known especially for his new pear varieties. Van Mons was originally a chemist. His teaching duties covered many aspects of agriculture. During the preceding 14 years, he had raised an immense number of seedlings and introduced new varieties of apple, pear, plums, cherries and peaches. He had taken to the habit of naming new varieties of pear after notable horticulturalists. Thus he had named a Sabine pear, after the secretary of the London Horticultural Society, a Knight after the president, a Salisbury after the eminent botanist and it seemed only fitting that he should name one of his latest varieties after Neill, as secretary of the Caledonian Horticultural Society. They could hardly wait to have a look round van Mons' fruit tree nursery, over the entrance door of which was the inscription 'Pepinière de la Fidelité'. They found the quality of the stock superior to anything they had seen so far. Van Mons was of the opinion that the best chance of obtaining good new pears was to use new varieties as parents, rather than the long established ones. He observed that he seldom failed to obtain a usable seedling from apple crosses. If the progeny was no good for dessert it could serve for the kitchen or for cider. Promising seedlings were often identified by their blunt shape, thinness and woolliness of the leaves and thin bark. Sometimes when a tree was cankered and had a new seedling grafted on it, the graft proved clean. Van Mons took this as evidence in support of Knight's theory of the degeneration of old stock. They were rather surprised to see that the nursery was freely accessible by the public, but fruit was not a scarcity here and it was neither pilfered nor damaged. It was agreed that van Mons would send a collection of grafts of apples and pears to Edinburgh. In April 1818 a large collection arrived, carefully packed in moss. Since the Experimental Garden was not yet available, they were taken care of by Messrs Dickson & Co. of Leith Walk and Messrs Dickson Brothers at Broughton. Dicksons' staff carried out the grafting so well that almost all the varieties survived to eventually find a place in the Experimental Garden.

While walking about Brussels, they encountered the Duke of Kent driving a curricle with a liveried servant by his side. Out of curiosity they went to have a look at where he stayed and were surprised to find it was an antiquated building that looked more like the office of an old manor house than a ducal palace. They expressed similar sentiments when they found that Lord Kinnaird had abandoned Rossie Priory in the Carse of Gowrie, for a dull, ruinous mansion in Brussels.

From Brussels they went via ill cultivated land of a rather impoverished peasantry, to Enghien where they met by the supervisor of the Duc d'Arenberg's estate. In the fruit garden they found the peaches represented by the White Magdalen, the Red Magdalen, Large Mignonne, Mignonne Double de Troyes and Large White nectarines. The plums included both the Green and Red Gages and the Swiss plum, which seemed the same as the Brussels Prune Altesse. They found a new variety of pear – the Duc d'Arenberg – of excellent quality. The foliage and wood were like that of the Winter Bon Chretien, and the fruit like that of the Brown Beurré, but never gritty at the core. This was indeed a variety for introduction. After this visit they went to the garden of a M. Parmentier who for more than 20 years had collected new plant species and introduced more than a thousand of them to the Low Countries, including plants grown from seed supplied by von Humboldt. His greenhouses and hot houses were full of exotics which included newly introduced species from South Africa and Australia. Both the Jardin des Plantes and the Berlin Botanical Garden had profited from M. Parmentier's zeal.

Their visit was marred by an extraordinarily intense storm which created general alarm. While being entertained by M. Parmentier, his servants were afraid to cross the courtyard to fetch wine for the guests so the host had to get it. The visitors ascertained that the most commonly grown pears included St Germain, Passe-colmar, the ordinary Colmar, Bon Chretien, Martin Sec, Virgouleuse, Rozid and Chaumontel. The best pears were grafted on Portugal quince stock, close to the ground. The quince spread its roots, unlike the tap rooted pear, and was therefore a very suitable stock for the shallow soil of the district. On their way to Paris they were impressed by the area of land devoted to growing white poppies, as well as colzat and rape. Oil extracted from poppy seeds was almost equal to olive oil. Coarser fractions from poppy and other species were used in the manufacture of cloth or woollens. Tobacco leaves were hanging out to dry outside many of the village houses.

At Paris they put up at the *Hotel de Boston*, near the centre. Their time in the capital was divided between their horticultural interest and sight-seeing, but Neill spent little time describing the sites, since he reckoned that most of his readers would know them better than he did – rather a significant aside that confirms the impression that there was a tremendous influx of British visitors to France and Belgium after the end of hostilities. In the fruit-market, pots of alpine strawberry caught their attention. The variety on sale here was different from the one sold at home. It had a narrower, more oblong shape and was called Majauf or strawberry Barremont. Like the Dutch gardeners, they discarded the plants every two years, but unlike them they raised new plants from seed.

Naturally they made for the Jardin des Plantes, which impressed them greatly. Neill provided a very detailed description of the lay-out and was glad to report that 'any intelligent naturalist, of whatever country, will here meet with every encouragement'.

They took a cabriolet to St Germain to meet Thomas Blaikie, who had left his native Edinburgh about 50 years previously to settle near Paris and pursue his career as 'ingenieur des jardins anglais'. They found him in the house of an English gardener by the name of

Hudson, who had been principal gardener to Josephine at Malmaison and at her death had retired to a pleasant spot where he cultivated his garden, partly for profit and partly for pleasure. Blaikie was no mean conversationalist and Neill and his companions listed with great interest as Blaikie recounted his experiences. In 1775 he had been employed by Dr Fothergill and Dr Pitcairn – they had notable plant collections in Upton in Essex and near London respectively – to collect alpine plants in Switzerland, for which he never had due credit. The records sent to Aiton's *Index Kewensis* were under either Fothergill's or Pitcairn's name and Blaikie got no mention. In fact, Blaikie can be regarded as the first alpine plant collector. By coincidence, Hay had spent a year in Fothergill's garden so he knew the background. After settling in France, Blaikie had a distinguished career designing gardens for the French nobility. His first task was to lay out a large garden for the Count Lauriguy at Mont Camissée. In 1778 he settled near Paris. There were no nurseries nearby and he had great difficulty obtaining trees and shrubs. In 1779 he worked at the princely villa of Bagatelle for the Compte d'Artois. M. Bellanger did the parterres and shrubberies near the house while Blaikie handled the extensive grounds, which took in part of the Bois de Boulogne. Blaikie had the garden paths laid with gravel, but found the gardeners unwilling to keep them tidy.

Ten years later, when Bagatelle was at its height of glory, the situation changed dramatically for it became the preferred site for fêtes during the revolutionary period. After 25 years it reverted to its original owner and although it had suffered, was still a beautiful place. Blaikie worked at Morceaux for the Duc d'Orleans, and was especially concerned with the erection of very fine hot houses. Neill observed that – as with his pioneering alpine collecting – Blaikie did not receive credit for these and other garden designs, since French authors deliberately failed to acknowledge his contributions. During the Revolution, he was commissioned to supply potatoes for the Tuillerie Gardens. He wanted to return to Scotland but was engaged in a protracted struggle to recover various debts owed to him. Blaikie – or 'the venerable Scotto-Gallican', as Neill termed him – was a truly remarkable person, whose career has been fully described in a recent biography.[2]

They next visited the gardens of the Palace of St Germain, accompanied by Blaikie who gave them a running commentary on the species and varieties of fruit trees. One novelty they learned about was the use of a preparation called Poudrette, which was used extensively as a fertiliser in kitchen gardens and incorporated in compost. This turned out to be dried night soil and (one suspects) Neill was a little taken aback to discover that it was sold in London as Clark's Desiccated Compost. It contained quicklime to mitigate the smell.

At the Tuileries they visited the Compte d'Escars, who immediately recognised Macdonald and spoke of his happy days at Dalkeith Park when he was in the suite of the Compte d'Artois at Holyroodhouse, during the Revolutionary period of exile. They found the Tuillerie Gardens still showing evidence of their designer Le Nôtre. At this point in his account, Neill was moved to reflect on some of the desirable features of Paris. For one thing, the air was remarkably clean, quite unlike the heavy pall of smoke that lay above London, created by innumerable coal-fires. In Paris they burned mostly wood and during summer did not use fires much for cooking, since families often dined at restaurants. In addition, he was struck by the way gardens of flowers and statues displayed in the public parks were not vandalised, even by ragged urchins who would have been very unpopular if they did any damage. By contrast, public walks and comparable places in Edinburgh were subject to all kinds of destruction such as 'the youthful mind could devise'.

He also liked the Parisian flower-market and wished there was something like that at home. The Caledonian Horticultural Society had already appealed to the provost and magistrates of Edinburgh to establish a fruit and flower market and he hoped something would come of it. Here fashion ruled. Every year some particular flower came into fashion and was in great demand until the bourgeoisie took it up, whereupon it lost its aristocratic cachet and began its decline in popularity, to be replaced by something different. The only exceptions to this cycle of popularity were the periwinkle, Neapolitan violet, Indian cress, scented heliotrope and mignonette. When the visitors passed through the Tivoli pleasure ground – which was not up to Vauxhall standards – they expressed surprise at not encountering a single drunk. Among the ruins of the Bastille they commented on the vigorous growth of half a dozen different wild flowers, which were identified by their scientific names. Neill went to a meeting at the *Palais des Beaux Arts* where he was introduced to a number of the savants noted for their knowledge of agriculture. He was agreeably surprised to see Mr Playfair – Professor of Natural Philosophy at Edinburgh – come into the meeting, where he was welcomed respectfully and led to a seat near the president.

They found the gardens at La Malmaison quite charming although rather neglected. During the time of Josephine, the flower garden was one of the richest in Europe. The magnificent glass houses engaged Hay's attention. He made a note of their dimensions and construction while Neill and Macdonald admired their contents, which included the original plant of *Brunswegia josephinia*, brought for the Empress 17 years previously from the Cape of Good Hope by a Dutch collector. They were given three or four ripe seeds, one of which germinated to produce a flourishing specimen at Dalkeith.

Naturally they visited Versailles, which was described in some detail. The ruins of the rustic cottage where the late queen had entertained visitors were still evident. After the Revolution an application to form a botanic garden had been approved by the National Academy. Blaikie had been offered the post of commissionaire but had declined. Neill noted as an aside, that the *Confréres de France* comprised the oldest gardeners' society in the country, just as Adam's Lodge in Aberdeen was the oldest in Scotland. They visited Vilmorin's Nursery and were overwhelmed by the immense collection of exotic plants but had to cavil at the lack of neatness. Edinburgh nurseries like those of Dickson & Co., Eagle and Anderson or the Dickson Brothers were incomparably superior in that respect.

The peach gardens of Montreuil were of special interest. A considerable area several miles across was covered with small enclosed gardens and innumerable dazzling white walls. Production for the Paris market was the business. Since different kinds of peaches required walls of different aspect, the number of walls proliferated, with individual gardens often with five or more of them. Their aspect was generally south east, south or south west, but sometimes they seemed to have been set down haphazardly. The late peaches needed maximum sun, so they were generally on south facing walls. The west aspect was also suited for cherries, plums and early grape vines. The subsidiary walls were 80 to 100 feet long and generally not more than 40 to 50 feet apart, 15 inches wide at the base and tapering slightly upwards. They were roughly built of small stones, with the mortar composed of the garden soil, and plastered on both sides with gypsum, which soon became hard enough to take a nail and created the white appearance. Along the borders on the north side of the fruit walls, filbert nuts were successfully grown.

One of the gardeners showed them around. The fruit trees were generally trained fanwise or 'en ouvert' with two main branches and two to three subsidiary branches. Branches and twigs were attached directly to the wall – without the treillage so commonly used in Scotland – by nails and shreds of woollen cloth called *loques* or *loquettes*. Although the trees were cleverly managed, they thought they were planted too close to the walls since the back of the stems was often flattened. They were informed that almond trees were used as stock for peaches, whereas in Scotland plum was generally used. However, their informant made the point that since the local soil was dry, almond was best, but for heavy, wetter soil, he also would use plum as stock. The rest of the day was spent looking round the village which was entirely devoted to fruit culture for the Paris market. The fruit was taken to Paris by women of all ages, starting between one and two o'clock of the morning, since the market was very early. There were no big firms in the village, which was made up entirely of family growers who lived very well, growing their own vegetables and making their own wine.

While waiting for permission to visit the royal gardens, they visited the shop of Vilmorin Andrieux & Co: the best place for seeds in Paris. They purchased a number of packets of culinary herbs which were subsequently cultivated by Macdonald at Dalkeith and Neill in his garden at Canonmills. Ever observant of their religious duties, they went to the Lutheran church, where they found that the texts for all the sermons for the rest of the year stuck up on the church walls. In Scotland, the subject of discourse was kept secret until announced from the pulpit by the minister.

The next garden on their list was that of Morceaux, which Blaikie had laid out in 1784 for the Duc d'Orleans. Although it had suffered neglect during the revolutionary years, it was still a fine place and had not entirely lost the character of the English style that Blaikie had established. They got wind of a grand dinner to be given to the diplomatic corps by the English ambassador, so they went along to the hotel to see what kind of fruit would be served up. The melons comprised cantaloupe Noir des Carmes and Melon de Malte. The peaches were excellent and included Teton de Venus, the Late Admirable, Chevreuse and Maltese. The nectarines were good, especially Brugnon Musque and Grosse Violette. The plums were chiefly St Catherine and the Late Damask (Damas de Septembre or Prune de Vacance). The pears were confined to the Red and the Autumn Bergamot and the Grey Beurré. The grapes included Chasselas de Fontainbleau, Musk Chasselas, Red and White Chasselas, but they were disappointing both in size and ripeness. This was generally their opinion about the grapes in French markets, compared with what they were used to. The most likely reason was that the French grapes were grown outside, whereas the ones on sale in Britain had been carefully nurtured under glass.

Neill went to the theatre to see Voltaire's *Oedipe*, which was well acted. He was somewhat taken aback by the audience's reaction. At the saddest scenes, the French gentleman next to him – who had explained various aspects of the play to him and lent him a copy of the play – was so overcome with tears of emotion that he could not respond when Neill turned to him with a comment. Many others in the audience were dabbing their eyes as well.

Since it was now late in September they were due to return home, without visiting a number of important nurseries that they had hoped to see. However in 1821 Neill independently visited the continent again. He made a point of visiting these and included his impressions in his account. Thus on August 17 1821 he visited the premises of the celebrated *pepinière* of Noisette at Faubourg St Jacques and was shown round by the owner. The nursery contained a very extensive collection of fruit trees suited to the local conditions. The species were arranged according to the classification system of Jussieu, while the varieties were arranged according to their time of maturity, following Duhamel's scheme. For each kind there was a specimen tree in bearing state and hard by, several young trees of the same sort, budded or grafted from the specimen tree. Most of the specimen trees were grown *en quenouille* so they did not crop heavily, but produced enough fruit to enable prospective purchasers to make sure of what they were getting. Neill thought this a capital idea, since it eliminated the chance of confusion about varieties and their names, and thought British growers could well copy it. The apples were grown on paradise and *doucin* stocks. The former was best for trees grown *en buisse* and *en quenouille* while the *doucin* was preferred for espaliers and half-standards. Both stocks were multiplied by layers and offsets.

Among apples, Noisette preferred the Summer Calville, for autumn the English Rennet, and for winter the Common Rennet, of which he sold ten times more than of any other kind. The Canadian, Golden and Grey rennets were the most esteemed for dessert, together with the White Spanish – which was a long cylindrical apple, covered in farina, which Neill thought might be suited for a west wall in Scotland. Noisette recommended the Reinette de Runeville and expressed a clear preference for the White over the Red Calville. Neill found the Pomme d'Api rated higher than expected. Small trees in pots were sent from Orleans to Paris in great numbers.

Among the esteemed pears was the Epargne or Grosse Cuisse Madame, which corresponded to the large or improved Jargonelle grown in Scotland. This was ripe one or two weeks before the ordinary Cuisse Madame, which in old gardens in Scotland, was often called the Jargonelle – such were the hazards of synonymy. The Rousselet de Rheims was a small, round, very juicy and perfumed pear. Another of some merit was the Poire d'Oeuf. The Autumn Crassane was regarded as the best, with Doyenne Gris coming next. Beurré Gris and Beurré d'Angleterre were treated as distinct varieties by Noisette. The so-called English Beurré appeared to be just a smaller variety of the Beurré Gris. Monsieur Jean and Calebasse were fine autumn pears. For winter, St Germain was rated foremost, followed by Passe-colmar, Beurré d'Aremberg, Winter Bon Chretien, Martin Sec, common Colmar and Virgouleuse. The Sylvange was unknown in Edinburgh and the Chaumontel was not thought highly of here. Neill suspected that the growers in Jersey knew better how to handle it, for they regularly sent to the London market large fruit with the stalks tipped with sealing wax, and these were quite free of the grittiness of which Noisette complained.

The peach collection was very good. It included Grosse Mignonne and Belle Bauce – a large well flavoured sub-variety of the former. Neill thought it should be introduced to the peach houses at home, where it was unknown. Early sorts included Petite Mignonne, Early Purple and Chrevreuse Hative. He thought the Red and the White Avant or Nutmeg peaches had nothing to recommend them other than their precocity, since they were ripe in July. The Admirable was esteemed. The French Royal of Noisette was a variety of it, but the Belle de Vitry was distinct and probably the same as what was known as the Late Admirable at home. The sort Neill knew as the Yellow Admirable was here called Peche d'Abricot. The Teton de Venus was regarded as one of the best of the late peaches, although it was not a very good bearer. The most productive of the better varieties was the Chevreuse Tardive. The Peach apricot was considered the best variety although the common apricot was the most popular. At Paris, apricots were generally grown as standards but they were also grown in a dwarf form and then were known as Bretons or Batardeaux. The fruit was small but of high flavour. The plums consisted chiefly of Reine Claude, Monsieur and St Catherine. For cherries, Noisette considered May Duke the best, followed by the Royal Duke. The Kentish cherry was known here as the Courte Queue de Montmorency or Gros Gobet and had the reputation of being a shy bearer. Noisette had several plants of the large leafed cherry which he had got from Poland in 1816. The leaves could be a foot long and half as wide, while the fruit were also very large, but in Paris at least, they soon fell off. Noisette grew the large fruited Amber raspberry with great success, as well as the Pale Red sort. He had imported from England gooseberries which were shunned by the French, who considered them only fit for sauce to accompany mackerel. Neill was also shown examples of peaches and grape vines being grown on the same wall to conserve space. Noisette expressed admiration for the accuracy of Duhamel du Morceau's *Traité des Arbres Fruitiers*, which Neill decided he must recommend for the library of the Caledonian Society in spite of its expense.

Accompanied by Blaikie, Neill visited the Chartraux gardens next to the Palace of Luxembourg on August 20 1821. These were established by Louis XIV. The religious fraternity had very fine gardens and paid particular attention to fruit selection. By granting privileges and immunities, the King encouraged them to devote a considerable proportion of their revenues to the collection of plants and grafts of the best kinds of fruit trees from throughout Europe, making use of their monastic contacts. At one time the gardens occupied 80 French acres, largely devoted to fruit trees. The monks were first class organisers. They made good profits from their orchards and nurseries and at the same time, promoted the horticultural interests of France. They raised seedlings from good varieties like the Sarrazin pear – formerly known as the Blessed pear since the Carthusians originally raised it. At the time of the suppression of the religious houses in 1791 the garden was threatened with destruction but largely escaped, due to the astuteness of the monastic superior. Later more enlightened attitudes prevailed and the garden was designated as a national nursery for fruit trees and a school for free instruction in horticulture.

Although the original acreage was diminished by the encroachment of the gardens of the Palace of Luxembourg, the gardens and orchards impressed the visitor. There were about 90 different varieties of apple. It is worth noting that both the Golden Pippin and Nonpareil (in spite of its name) were regarded as of English origin. Two varieties of apple had been raised here, called respectively Belle Hervy (after the superior) and Concombre des Chartreaux. There were more than 190 varieties of pear. It turned out that what Neill knew as the large Jargonelle was here known by four names: Grosse Cuisse Madame; Epargne; St Sampson and Beau Present; so that cleared up the confusion about names at Ghent and Brussels. There were 16 sorts of apricot, more than 20 peaches and cherries and more than 70 kinds of plum. Before he left he was assured that a request for plants or grafts by the Caledonian Horticultural Society would be treated in exactly the same way as requests from comparable French institutions.

Neill called in at a garden which had been first designed for the Duc de Praslin by Le Nôtre, before he had attracted royal notice. For the preceding 30 years the kitchen garden had been in the charge of a Scots gardener called Macmaster, who had been introduced to the French scene by Blaikie. He had recently retired but retained a lively interest in Scottish affairs.

Calling in again at the Paris fruit market, to see what was on offer in August, Neill found the only really good pear available was the large English Jargonelle or Epargne. Although Windsors were on sale they were not ripe. Also on sale, were the Green Chisel, and the Skinless – both summer pears. There were also the Ognolet – unknown in Edinburgh and so-called because the fruit grew in bunches like onions – the Large Blanquet, the Long Stalked Blanquet, the common Cuisse Madame, Poire des Deux Têtes (a large, sweet juicy pear without much flavour), the Caillot (with a flat fruit), the Musk Orange (of yellowish colour) and the Red Orange (which really was orange in colour). The Robine or Royale d'Eté was quite plentiful. The small early Rousselet was very common and cheap. Towards the end of August the Cassolette (a good, small, storing pear) was on sale, together with the Rousselet de Rheims. Hawkers in the streets were selling what they called the Poire d'Angleterre, which was a kind of Beurré pear.

The commonest apples included the Dutch and the Carlisle Codlin, Jenneting, Summer Pearmain, Hawthornden, the Summer Calville (a small conical dark red apple), and the Pigeonnet. Plums were present in quite a variety, especially the Green Gage, Prune Royale, Jaune Hative, Drap d'Or, Mirabelle, musk damask, Précoce de Tours, Prune Noire de Montreuil and Blue Perdrigon. The variety of plums at the Paris market greatly exceeded what Neill had seen at Covent Garden in August, where the White Primordian and the Morocco were the only ones on sale. Apricots were also more abundant. They included the common apricot, the Portugal, Angoumois and Peach apricot.

Neill's account now picks up again on September 30 1817, when the three travellers were on their way home through the valley of Montmorency – which was real orchard country, with a reputation for cider and perry. Piles of apples lay by the wayside awaiting collection. As they jogged along, they saw people shaking down the fruit with the aid of long sticks. In one orchard on the way to Rouen, they saw old fruit trees which had undergone *dechenillence* – i.e. paring away of the old bark to get rid of insects. At Rouen Botanical Garden they thought the greenhouses were in a very clumsy style, with huge beams supporting the roof. They also visited the garden of a M. Vallet where there was a magnificent collection of orange trees, only exceeded by the royal collection. One of them was about 400 and several at least 200 years old, with many more large and small trees. The owner made quite a profit from the orange flower water which he produced by distillation, and by selling trees – many of them to English buyers. The visitors also learned that an English company Colvert had established a rose nursery at Bonne Nouvelle near Rouen, where they offered for sale nearly 900 varieties, although very few of Austin's numerous Scots roses were included. Neill observed very sensibly that, without disparaging the zeal of the growers, it was a pity the numbers were not reduced to a tenth of those on offer and these should be distinguished by clearly recognisable differences. There was a considerable local trade in exporting pears to England, especially varieties like Crassane, Colmar, St Germain, Chaumontel and Bon Chretien.

When they got to Dieppe on October 3rd, they found a film of ice on the wayside pools. The harbour was tidal and poorly constructed. Several hundred women with baskets were engaged in the hopeless task of clearing the channel by removing gravel from one side to the other, rather pointless since the next gale would undo all their work. About five in the evening, with the aid of about 150 of the women, their schooner was hauled down the channel to the sea and they were on their way to England.

They spent several days in London. Hay went off to reconnoitre at Portsmouth and Macdonald went to the Duchess of Buccleuch's seat at Richmond, while Neill went to visit his old friend James Dickson – the distinguished cryptogamic botanist, then in his 80th year. Since the council of the London Horticultural Society was meeting that day, Dickson proposed that Neill should accompany him to the session at one o'clock at the Linnaean Society's rooms in Gerard Street. Joseph Sabine was in the Chair. Specimens of uncommon species and varieties of plants were displayed for comment and discussion and fruits were tasted. Members often brought specimens in their pockets for identification. Sets of a new strawberry (Wilmot's Scarlet) were presented to those who were willing to grow them as a trial – including Neill – as a representative of a sister organisation. In all, Neill thought the proceedings entirely filled his expectations. James Dickson died in 1822 and as Neill put it: 'feeling the ruling passion strong in death was, by his own desire, buried in a romantic churchyard among the Surrey hills, where in his earlier years, he had been accustomed to gather rare mosses.'

After Hay returned from his trip, Neill was unwell so their return home was delayed a few days. Hay and Macdonald carried on with their visits to nurseries, beginning with the celebrated establishment at Hammersmith of Lee and Kennedy, where they saw a rich collection of cape heaths, camellias and pelargoniums. At the Duke of Devonshire's seat at Chiswick they admired the magnificent set of hot-houses. At Kew they saw a recently imported plant of what was called *Cactus cochinillifer*, still bearing live cochineal insects. At Spring-Grove – the seat of Sir Joseph Banks – they saw the pond, round the margin of which the large fruited cranberry was first raised. At Sion Hall the pineapples were equal to anything they had ever seen. Armed with a letter from Lord Montgomery to General Taylor – secretary to the King – they were given unrestricted entry to the gardens at Windsor and then on to Frogmore – a residence of Queen Charlotte – where they were given a fruit of *Passiflora edilis* with ripe seeds. The next day after breakfasting with Lee at the nursery, they visited the Bishop of London's garden at Fulham to see some of the trees planted originally by Bishop Compton. At Mr Angerstein's garden they were presented by Mr Macintosh – superintendent of the pleasure grounds – with a cutting of the old Frankendale vine, which Macdonald was glad to have. At Mr Grange's fruit and market garden at Hoxton, they were invited to breakfast before surveying the 57 acres of grounds, where they were impressed by the very extensive hot-houses and the large ice-house which supplied ice for his elegant fruit shop at Westminster. After dining with Mr Grange they inspected an immense collection of pelargoniums arranged in separate houses, as well as cape heaths and camellias at Smith's of Dalston – where many plants were grown in pits and frames banked up with earth to protect from frost. This marked the end of their tour of inspection. Since Neill had recovered, they immediately set off for Edinburgh, which they reached in a few days time.

It may be wondered what were the final impressions and conclusions of the three horticulturalists at the end of their continental foray. Neill summed them up in the preface to his book, but they are more appropriately recorded here. To quote: 'It may perhaps be thought that we have announced very few improvements in the general style of gardening or even in particular practices of culture as existing in the foreign districts which we visited. The truth is, we were lead to form the opinion that our own style of gardening in Scotland is, generally speaking, superior to what we witnessed on the Continent; it may be very true that we originally derived our horticulture from the Flemings and the Hollanders, but it seems equally certain that we have now, in many respects, surpassed them.' So, on the one hand they may have been a little disappointed at not encountering novel practices which they could adopt, but on the other, they could feel with quiet satisfaction, that they were well placed to establish and administer the Experimental Garden, which was the next major objective of their society. They had made many useful contacts on their trip with nurserymen and fruit breeders, who would in future supply them with seeds and grafts of fruit trees, and at the same time they had brought the Caledonian Horticultural Society to the notice of a great many foreign horticulturalists, both commercial and academic.

---

1 Patrick Neill, *Journal of a Horticultural Tour in 1817 Through Some Parts of Flanders, Holland and the North of France in the Autumn of 1817 by a Deputation of the Caledonian Horticultural Society*, Edinburgh, 1823, printed for Bell & Bradfoote and for Longman, Hurst, Rees, Orme and Brown, London.

2 Patricia Taylor, *Thomas Blaikie (1751–1838) The Capability Brown of France*, 2001, Tuckwell Press, East Linton, Scotland.

9 ❧ *Neill's Friends*

Neill had a wide circle of friends and acquaintances drawn from all levels of society. As a leading printer, executing university and government business, he was from an early age in his career brought into daily contact with Edinburgh's scientists and scholars who would meet informally on his premises in Old Fishmarket Close.[1] Although his own university career and the prospect of entering the medical profession had been cut short by the demands of the business, Neill was entirely at ease in the academic fraternity. As secretary of both the Wernerian Natural History Society and the Caledonian Horticultural Society, he was as well acquainted with the scientific community of Scotland as with the owners and head gardeners of many estates and the local nurserymen.

Although the loss of his papers means that we can never know the extent of his daily contacts with such kindred spirits, we can confidently identify his principal friends from correspondence or other sources. Many of them lived in Edinburgh. Others living elsewhere kept up a regular correspondence, generally about plants. Among the former, the Professors of Botany and the successive superintendents of the Edinburgh Botanic Garden and the Experimental Garden of the Caledonian Horticultural Society were particularly close. As a devoted gardener and botanist who spent much time and money on his own garden, the Edinburgh Botanic Garden was a source of endless interest to Neill. When the Garden was transferred from Leith to Inverleith in the early 1820s it was only a few minutes walk away from Canonmills Cottage, so he was a very regular visitor, keeping an eye on what was flowering and often campaigning in print for better government funding for both staff and the care and maintenance of the greenhouses.

The first of the Botanic Garden superintendents who was a close friend of Neill's was **John Mackay** (1772–1802).[2][3] A native of Kirkcaldy, Mackay was the son of a professional gardener who transferred his family to Inveresk when John was aged 14, by which time he was already an avid plant collector. At 18 he went to work at Dickson's nursery at Leith and soon made his mark. Dickson's took a relaxed view of their unusually knowledgeable employee and allowed him to go on long visits to collect plants in the Highlands. He became particularly interested in alpines. During the summer months he also helped out at the Botanic Garden. It was not long before Neill and he became friends and companions in plant hunting excursions about Edinburgh. Neill wrote an account of one such trip with Mackay in April 1800, not long after Mackay had been appointed superintendent of the Botanic Garden. They sailed to the islands of Inchkeith and Inchcolm in the Firth of Forth, where they found a number of interesting plants. This trip was described in his *Antiquaries Journal* and has been dealt with under that heading. Tragically, Mackay died in 1802. Neill wrote a moving obituary of his friend in *The Scots Magazine*, expressing the conviction that had he lived, John Mackay would have risen to great distinction in his profession. That was no exaggeration for Mackay had already been elected an associate of the Linnaean Society and had impressed the leading botanist James Edward Smith, who acknowledged his communications about many different species of British plants.

John Mackay had a younger brother – **James Townsend Mackay** (1772–1862) – who realised the sort of distinction John might have won.[4] In 1804 he went to Dublin as assistant botanist and two years later, was put in charge of the garden newly-established by the Board of Trinity College. Here, lecturing and demonstrating to medical students, he developed the site into the well-furnished Botanic Garden and is duly honoured for his long tenure as the first curator. James Mackay was, like his brother, an enthusiastic field botanist who soon became an authority on the Irish Flora. He made a point of visiting other gardens in Britain and was often in Edinburgh where he stayed with Neill at Canonmills. The two men were very good friends. Neill remembered him in his will in the warmest terms.

John Mackay was also a friend and botanising companion of **George Don** (1764–1814) who shared his passion for plant collecting. George Don was a gardener and nurseryman of extraordinary determination who established a remarkably diverse collection of British plants at his nursery of Doo Hillock in Forfar.[5] [6] There is doubt about the exact date of birth and earliest years, but apparently he was apprenticed to a clockmaker in Dunblane, where he started making a collection of dried plants. From there he moved to Dupplin Garden where a relative was in charge, and stayed there to complete his training and use his spare time in roaming the Ochils and beyond in search of plants. He then moved south to Worcestershire and later in 1786 or 1789, he was employed as gardener in a London nursery. A few years later, in or about 1797, he returned north, settling at Forfar with his modest savings to lease two acres of land on condition he build a cottage of specified size within a minimum period. This was Dove or Doo Hillock, which sloped down to Forfar Loch.

By this time he had begun to establish a reputation as a botanist, for he corresponded with some of the leaders in the field. When the post of principal gardener of the Edinburgh Botanic Garden became vacant – as a result the death of his friend John Mackay – on the recommendations of Sir J. E. Smith and Brodie of Brodie (who was also a botanist), Don was appointed.[7] Once in Edinburgh, he was able to cultivate his friendship with Neill, who accompanied him on botanising trips in the district. They had already met in Forfar and won mutual respect. Just how they met is worth quoting, especially as it reveals Neill in one of his favourite pastimes.[8] Thus: 'When on a pedestrian expedition along the east coast of Scotland I happened to spend a night at Montrose, and it occurred to me that both Brechin and Forfar deserved to be visited – the former for its well known Den and its round tower of remote antiquity and the latter for the remarkable garden, and its owner, whose fame was familiar to me owing to my intimacy with his regular correspondent Mr John Mackay of the Leith Walk Nurseries. In passing along the margin of the sea basin above Montrose, the bay being at ebb tide, I picked up some fine plants of *Salicornia herbacea*, then in flower and also a somewhat shrubby cicely. On reaching Forfar towards evening I soon found Don's garden and, entering, enquired of a very rough looking person with a spade, whom I took for a workman, whether Mr Don was at home. The answer was 'Why, sir, I am all that you will get for him.' Having apologised in the best way I could I stated that when I left home I did not anticipate to visit Forfar, else I could have brought a note of introduction from Mr John Mackay. Mr Don, pointing to my botanical box, said, 'That is introduction enough for me', and having inspected the contents, remarked that he was in want of an example of *Mondria mongyna*, an Equisetum not having succeeded, forthwith conducted me to the Linnaean arrangement. I was then introduced to Caroline, his wife who had brought him two sons and a daughter. I persuaded him to accompany me to the inn at Forfar, where he spent the evening with me. Next morning at six he met me then by appointment and conducted me to Restennet Moss, where I had the great satisfaction of procuring a large patch of *Eriophorum alpinum* and a number of fine specimens for drying. The Moss was at that time partially drained, for the sake of a rich deposit of marl, but at one end there was still sufficient marsh for the growth of *Schoenus (Cladium) mariscus* and *Eriophorum angustifolium* and, of course, for the rare *E. alpinum* which grows in the drier and firmer part of the Moss. Mr Don remarked that in a few years the plant would disappear, which I understand has accordingly happened.'

In a summary in the Wernerian Society Memoirs, Don published a list of many herbaceous plants and cryptograms not previously recorded from the King's Park, Edinburgh, and in 1804 began planting a Herbarium Botannicum, dedicated to Sir Joseph Banks. His time at the Botanic Garden proved rather unhappy. He did not get on well with the Professor of Botany and was rather uneasy with greenhouses. Neill later shed light on his time in Edinburgh when he published an obituary of Don.[9] Neill had to concede that he did not have experience in the cultivation of stove plants and hardly shone in that department. But as a botanist his knowledge greatly exceeded that of the professor, who though an accomplished chemist, had little time for botany and this contributed to the lack of cordiality between the two men. During his time in Edinburgh he followed nearly all the medical classes with the intention of pursuing that profession. However, the desire to lead a more untrammelled existence together with the wholly inadequate salary persuaded him to return to Doo Hillock, probably about the end of 1806. Between then and his untimely death in 1814, Don tried rather unsuccessfully to earn a living as a nurseryman and expand the diversity of his garden, which contained an immense number of species of British plants – including many alpines which Don collected on long expeditions into the Scottish hills, some of them new to the British flora. However his garden bore little resemblance to what visitors expected to find, for it was chaotic in the way the plants were distributed in his plots. Nonetheless, Neill was full of admiration for his field knowledge and was sure he was the most eminent practical botanist in the country.

Although Neill never wavered in his admiration for Don's botanical zeal, in due course he had to recognise shortcomings in his attitude to other botanists, including himself. Thus in November 1804 he received a letter from J.E. Smith – founder of the Linnean Society and joint author with Sowerby of the celebrated *English Flora,* in which Don complained of plagiarism on the part of Neill's friend John Mackay with respect to the credit for finding particular plants.[10] Sadly Mackay had died a year back. Neill replied that he had never heard such complaints during Mackay's life. When he raised this delicate issue with Don the latter claimed that living in the country, he had no chance of reading the publications in which Mackay's accounts appeared. Neill thought this credible but hardly adequate. With the greatest respect to Don, Neill had to recognise that Don had been too eager in 'appropriating his discoveries' and to have fallen into a habitual inclination to detract from Mackay's merits, sometimes 'in a very rough manner'. Neill had remonstrated with Don on this score on several occasions. When Don recorded finding a particular plant at a particular site on a particular date, his claim could be accepted, but when he referred to the finding of species when he was not present 'one should hesitate'.

As far as Neill could recall the two cases of alleged plagiarism on the part of Mackay referred to *Stellaria cerastoides* (*Cerastoides caespitosa*) and *Thalapsi hirsutum.* Mackay had reported that the former occurred north of Invercauld, Braemar. Don asserted that Mackay had never been so far north. But Neill, anxious to preserve the good name of his friend, observed that, to record the occurrence of a species at a particular place did not necessarily imply that the author had found it there. As for the *Thalapsi,* Don claimed that Mackay had got it from the garden of the Kinnoul estate, where the gardener had found it in the district. However, Neill thought that 'silly vanity' would never have impelled Mackay to stray from the truth. Smith dealt with this awkward issue by attributing the finding to Don and other botanists without mentioning Mackay.

Neill himself did not escape Don's criticism. In May 1806 Smith wrote that Don had complained that Neill had claimed finding five species in a recent publication.[11] This referred to Stark's *A Picture of Edinburgh,* noted earlier. Neill replied that his appendix was not intended as a scientific publication but only a popular sketch of what was to be found about Edinburgh. He had intended it to be anonymous but Stark had independently attached his name with an embarrassingly flattering introduction. In a second edition he could make it clear that he laid no claim to discovering the plants and so forestall Don's 'over-jealous claim'. In any case, two of the species – *Hieracium umbellatum* and *Valeriana pyrenaica* – had already appeared in Don's *Herbarium. Poa diastans* had already been referred to in a review of Sibbald's list, published in *The Scots Magazine* for 1802. The other two species – an *Eriophorum* and a *Galium* – could hardly be called rare Scottish plants and a botanist of Don's standing could hardly claim discovery.

Neill then observed that Don was 'a most acute botanist' but Smith must be aware of how confused and inaccurate his writing was. Neill had been put to a great deal of trouble in extracting the habitats and comments in Don's five published fasiculi. They were still not right but would have been infinitely worse had he not transcribed the greater part of the text. But perhaps much delicacy of feeling was not to be expected in one who had few opportunities for improvement. Don had never raised his criticisms to his face but had spoken to others of the 'robbing of his discoveries'. Such conduct increasingly convinced Neill that he was right when he had originally doubted the validity of Don's attacks on Mackay.

When he died unexpectedly he left his widow with three sons and a handicapped daughter, entirely penniless. Neill quickly moved to raise money for their immediate wants by writing to leading botanists like Smith and Hooker, to persuade them to encourage members of the Linnaean Society – of which Don was an associate – to contribute to a fund.[12] At least £80 was raised, which Neill administered with scrupulous attention. He enlisted the co-operation of several people who lived in Newburgh, to which Mrs Don had moved.[13] They could visit the distressed widow and decide how best to help. Thus on August 14 1815, he wrote to David Booth of Newburgh, reporting that he had recently received a letter from Mrs Don via a local minister who was applying for money on her behalf. He had written to the minister explaining the facts relating to the £80 from Sir William Hooker and expressing the hope that Booth and several other gentlemen could be consulted on how it should be dispensed.[14] He had declined drawing any funds from the bank until he had received a letter signed by at least two of the responsible persons requesting him to do so. In the spring of the following year he had such a joint request to send £10 to Mr Booth, to be judicially applied by him on behalf of Don's sons in Newburgh. He asked Mr Booth to ascertain their needs and order such clothes as he thought necessary.

Neill's association with the Don family did not end there. The eldest son **David Don** (1799–1841) became a gardener, working first at the Dickson nursery and then at the Edinburgh Botanic Garden, from where he advanced to become a distinguished taxonomist and Professor of Botany at King's College, London.[15] He kept in regular touch with Neill and stayed with him from time to time at Canonmills. George Don's other surviving son – also named **George** (1798–1856) – was a botanist too. He became foreman of the Chelsea Physic Garden and distinguished himself by collecting plants in South America and Africa for the London Horticultural Society.[16]

Neill's other friends associated with the Edinburgh Botanic Garden included the successive Professors of Botany **Robert Graham** (1786–1845) and **John Hutton Balfour** (1808–1884), and the successive superintendents of the Botanic Garden, **William McNab** (1780–1848) and his son **James McNab** (1810–1878).

Robert Graham was appointed the first Professor of Botany at Glasgow University in 1818 and two years later, succeeded to the Chair of Botany at Edinburgh University and Regius Keeper of the Botanic Garden – posts he held for the rest of his life.[17] During his tenure at Edinburgh, Graham and Neill were in frequent contact as Neill was always in and out of the Botanic Garden. Graham made a habit of publishing records of plants which had flowered for the first time in Scotland. Naturally most of these were grown in the Garden greenhouses or stove but a substantial number were also recorded from Neill's collection, which provided specimens which were figured in botanical journals.

In 1845 Neill was canvassing on behalf of Joseph Hooker who was competing with John Hutton Balfour for the Edinburgh Chair of Botany. Balfour was appointed but Neill's backing of his rival did not affect the cordial relations which were established after his appointment.[18]

For the greater part of Graham's term of office he had the benefit of William McNab's management of the Botanic Garden. Neill never lost an opportunity to speak highly of his skill. William McNab was a native of Ayrshire. During the early years of the 19th Century he served as gardener at Kew but in 1810, on the recommendation of Sir Joseph Banks, he was offered the post of superintendent at Edinburgh. In spite of the much lower salary, he accepted. The story of his enterprise in developing the Garden is well known.[19] When he was appointed it was located in Leith, but in the early 1820s the contents of the garden were transferred to a larger site at Inverleith. McNab's prowess in transporting large trees through the streets of Edinburgh to the amazement and acclaim of the inhabitants has become a horticultural legend.

His son James followed him in 1848, but before that he was superintendent of the Caledonian Horticultural Society's Experimental Garden, to which he was appointed in 1836.[20] He had established a reputation as botanist during his plant hunting expedition to America the previous year. During his service with the Horticultural Society, James McNab displayed remarkable drive and dedication in establishing new conservatories, a camellia house, a propagation house, a Winter Garden and an exhibition hall.[21] To his skills as gardener he added a flair for administration and the art of raising money. When he moved to the Botanic Garden he worked closely with Professor Balfour in the design and heating of impressive new greenhouses, in which he became an expert. James McNab was a particularly close friend of Neill, who as secretary of the Horticultural Society worked closely with him in the establishment of the Experimental Garden.

The regard between the two men was mutual. It was James McNab who organised the remarkable testimonial and silver vase to which 600 Scottish gardeners subscribed in honour of Neill in 1843. McNab was one of the three executors of Neill's will. He joined with the other two executors in an approach to Professor Balfour, expressing the hope that he would prepare an account of Neill's life and achievements using his journal and papers, which they would make available for that purpose.[22] But Balfour was not given to writing about anything other than scientific botany and to the loss of posterity nothing came of this appeal.

Robert Kaye Greville (1794–1866) was another botanist friend of Neill. A native of Bishop Auckland, Greville became a man of independent means early in life and settled in Edinburgh where he devoted himself to botany. He became expert on mosses, liverworts and algae.[23] He was an excellent artist, illustrating his own publications. He provided the sketch of Neill's garden, which Loudon published in his journal. He was commissioned by Professor Balfour to prepare botanical drawings and illustrations for use in teaching. His *Scottish Cryptogamic Flora*, based on his many botanising expeditions about Scotland, stands as a major contribution.

John Fleming (1785–1857) was a particularly close friend. A native of Bathgate near Edinburgh, he graduated as Doctor of Divinty at St Andrews and became a minister of the Church of Scotland, serving in the parishes of Bressay, Flisk and Clackmannon. However his interests in natural science led to his appointment as lecturer in natural history at Cork and later he was appointed Professor of Natural Philosophy at Aberdeen University. At the Disruption of 1843, siding with the dissenters, he was later appointed Professor of Natural Science at the New College (Free Church) Edinburgh.[24] Fleming was determined to combine his devotion to theology with an academic career. He was quite upset when Neill failed to let him know that the Chair of Agriculture at Edinburgh University had become vacant. But since the salary he could expect was only half of what he was then receiving in his ministry, Neill – not unreasonably – thought that the interests of his family would hardly favour such a change of job. Fleming won respectable status with his books *The Philosophy of Zoology* and *History of British Animals* and was known for his efforts to reconcile biblical teaching with scientific evidence. Neill named him as one of his executors and – as already noted – bequeathed to him all his unpublished manuscripts and papers. Fleming declined to act as executor and died six years after Neill.

John Gillies (1792–1834) was a native of Edinburgh and a naval surgeon who in 1820 went to Buenos Aires for health reasons, and spent the next nine or so years in South America where he collected plants in both Argentina and Chile.[25] He built up a collection of dried plants and sent roots, bulbs and seeds to his correspondents in Britain, including Neill. Several species, including *Tropaeolum pentaphyllum*, were first flowered in Britain in Neill's greenhouse from specimens sent by Gillies. In 1829 he returned to Edinburgh but lived only a further five years. Reflecting sadly on his early death, Neill surmised that it was hastened by Gillies' frustration at the turn of political events, which he took seriously.

Another friend associated with South America was **John Tweedie** (1775–1862).[26] Tweedie, from Lanarkshire, was a gardener who worked early in his career at Dalkeith and the Edinburgh Botanic Garden. Later he went to Castle-hill near Ayr, then Sundrum, nearby, then Blairquhan Castle and finally to Eglinton Castle. At the age of 50 he was tempted to leave his congenial post at Eglinton and head for South America, presumably tempted by the botanical riches he expected to find there. He spent the rest of his life in South America, during the course of which he collected plants from Bahia Blanca in the south to Tucana in the north. He kept up a regular correspondence with botanists in Britain, including William Jackson Hooker and Neill, and sent back live specimens – some of which were received and successfully cultivated by the latter.

Robert Jameson (1774–1854) was in a rather different category from these botanist friends.[27] A native of Edinburgh, he originally intended to follow a medical career but was encouraged by the Professor of Natural History to pursue his interests in science. He was given the task of looking after the Natural History Museum – an undertaking which became a life-long responsibility and which grew under his care to become so immense that, after his death, it became the foundation of the Royal Scottish Museum. In 1803 Jameson became Professor of Natural History at Edinburgh University and held the post until his death. The term 'natural history' was used more broadly in those days and included geology, which together with mineralogy, was Jameson's prime interest. Jameson established a great reputation. He was member or corresponding member of a great many scientific societies, both in Europe and America. He founded the prestigious Wernerian Natural

History Society and filled the post of president throughout its career. The significance of the term 'Wernerian' and why he chose it is has already been considered. Neill – who was a co-founder of the Wernerian – worked closely with Jameson in conducting the business of the society. Neill felt a great respect for Jameson and his scientific achievements. So it is hardly surprising that he leapt to his defence when the young Henry Hulme Cheek had the temerity to pour scorn on Jameson and his domination of the Wernerian Society in the pages of the short lived *Edinburgh Journal of Natural and Geographical Science*.

George Combe (1788–1858) was an Edinburgh lawyer who became a leading phrenologist and social commentator.[28] He developed a secular and rational philosophy of human society, which won sympathy from progressive and liberal thinkers and fierce condemnation from ministers of the Church. Neill appears to have become involved with Combe and his ideas when he was doing printing work for him. He also supported Combe in his unsuccessful attempt to be elected to the Chair of Logic at the University of Edinburgh. Phrenology – or the science of the mind – was the name given to a theory that the brain was made up of organs or parts that differed in function. The size of such regions indicated the relative strength of their influence on behaviour. Such differences in size were reflected in the dimensions of the cranium so that careful measurement of the skull could indicate differences in behaviour. Neill became converted to phrenology and was in favour of its practical application. In 1836 he wrote to Lord Glenelg – Secretary for the Colonies and responsible for supervising the transportation of convicts to Australia.[29]

He prefaced his letter with a brief explanation of the theory of phrenology and how it could provide a guide to behaviour. In the case of convicts to be transported, a phrenologist could immediately identify those individuals who were most likely to prove treacherous and malicious during the voyage, so they could be more securely guarded and kept apart from those who were identified as benevolent. On arrival the former should be kept at work under public surveillance while the latter could be hired out to settlers. To show his competence in making such recommendations, he mentioned that he had spent three years at university with the intention of entering the medical profession and that he witnessed the dissection of the brain by both Monro Scecundus and Sparzheim – a founder of phrenology. There is no evidence that his suggestion led to any change in either the classification of convicts or how they were treated.

In spite of their agreement on the value and importance of phrenology, Neill and Combe differed on many other issues. During the 1830s and 1840s they indulged in a protracted correspondence in which Neill defended the Calvinist position and the tenets of the Church of Scotland in the face of Combe's criticisms. But they never could agree. Although they differed in theological outlook, this did not diminish their mutual regard. In this respect Neill differed from many of his co-religionists who castigated Combe as a sort of Anti-Christ.

Among the other Edinburgh scientists who, if not such close friends of Neill were well known to him was **Sir David Brewster** (1781–1868).[30] A native of Jedburgh, Brewster was originally intended for the Church and completed his divinity degree at Edinburgh University and was licensed to preach. However his interests in physical science took over and he entered a distinguished career in the study of polarization, crystallography and the laws of refraction and absorption of light, which won him membership of the Royal Society of London in 1815. He was a prolific editor and publisher of scientific journals, including *The Edinburgh Philosophical Journal* (jointly edited with Jameson), and later – after parting from Jameson – *The Edinburgh Journal of Science*. In 1807 he took on the editorship of *The Edinburgh Encyclopaedia*, which continued to be issued in parts until 1830. Neill wrote the considerable article on *Horticulture* in Brewster's *Encyclopaedia* as well as the shorter account on *Fuci*.

Other friends of many years standing included **Robert Bald** – mining engineer at Alloa, who kept Neill informed about the appearance of whales in the Forth – and also his almost exact contemporary **Robert Stevenson** – engineer to the Northern Lighthouse Board and grandfather of Robert Louis Stevenson. Neill left his daughter a substantial sum in his will. With the exception of Tweedie, all these men lived in or near Edinburgh. Neill often met them in business, on social occasions and entertained them at home, so communication was direct and unrecorded.

He had several other very influential friends who lived elsewhere in Scotland or in England – almost entirely botanists – and these we know about either through incidental comment or through letters which Neill wrote to them and which have not suffered the fate of the bulk of his correspondence. Among these none stood higher in Neill's estimation than **Robert Brown** (1773–1858), a native of Montrose.[31] His career ran a rather unusual course. After education at Marischal College in Aberdeen and then a student of medicine at Edinburgh University, he joined the army as surgeon's mate. A keen naturalist, he lost interest in a military career and came to the attention of Sir Joseph Banks, who was responsible for his appointment as naturalist to Michael Flinders' Australian Expedition. This gave Brown the chance to display his talents as botanist and plant collector, realised in his celebrated *Flora of Australia and Tasmania*, published in Latin in 1810. After his voyage he was appointed librarian to Banks and when the latter died, an arrangement with Banks' trustees transferred the huge collection to a new department of the British Museum under Brown's direction. He remained there the rest of his life. Robert Brown was regarded as one of the most eminent botanists of his day. He was the 'Botannicorum facile princeps' of von Humboldt. He was offered two Chairs of Botany – including Edinburgh, which Neill encouraged him to accept – but he declined them since he was not interested in teaching medical students. Neill appears to have been often in contact with Brown who visited him at home. On one occasion when Neill was on a visit to London, he discovered that Brown was about to sail for Edinburgh so he curtailed his stay and secured the berth next to his, so that he could savour his conversation during the journey north.[32]

Another botanist native of Edinburgh but later living in Glasgow was **George Arnott Walker-Arnott** (1799–1866), who succeeded John Hutton Balfour as Professor of Botany at Glasgow University.[33] He collaborated with Robert Wight – an East Lothian naval surgeon who spent 30 years in India, during which time he amassed a huge collection of botanical specimens. He brought them back in 1831 to provide Arnott with valuable material for study and naming. Arnott appears to have been another frequent visitor to Canonmills cottage. When George Bentham was visiting Scotland in 1823 and 1827 he left a record of his meeting with Neill who entertained him to dinner in the company of George Arnott on both occasions, as previously noted.

Three other friends require more attention than those noted above, since a significant proportion of Neill's letters to them have survived and thus provide a more personal and revealing glimpse of the relations between the correspondents. Chief among these was **William Jackson Hooker** (1785–1865) who was a friend and correspondent of more than 30 years standing.[34] William Hooker – son of a merchant's clerk in Norwich – demonstrated his attachment to botany from an early age, for when only 20 he had discovered a species of moss new to the British flora. He soon became an authority on the cryptogams, benefiting initially from the guidance of Dawson Turner – a Yarmouth banker and amateur expert on these plants, whose daughter he married. He also benefited from his friendship with another local botanist – James Edward Smith – who had purchased the herbarium, library and papers of Linnaeus. It was not long before he came to the notice of Sir Joseph Banks, who had a flair for spotting budding naturalists of promising ability. He advised him to undertake a botanic expedition to Iceland. Unfortunately on the way back the ship caught fire and his collection was lost. Hooker, short of cash, entered on his long career of publishing botanical works, four of which appeared by 1820 when he was aged 35. Hooker was appointed by the University of Glasgow to the Chair of Botany as a result of Banks' recommendation, which was decisive in so many scientific appointments. At first Hooker's appointment was very unpopular among the members of the Medical Faculty.[35] Traditionally botany was seen as almost a branch of medicine and they looked askance at a non-medical professor. However, Hooker settled happily enough in Glasgow, raised a family, botanised extensively in the Highlands, published *Flora Scotica* in 1821, successive volumes of *Exotic Flora* between 1822 and 1827, and *Flora Boreali Americana* between 1829 and 1840. In 1841 he left Scotland to take up the first directorship of Kew Gardens, where he remained for the rest of his life, publishing botanical works, developing the Gardens and establishing his reputation as the most influential British botanist of the day.

His friendship with Neill probably arose through the latter's responsibility for printing *Exotic Flora*. Thus in 1822 Hooker wrote to Neill to make a case for adopting a page size that would not 'clash' so much with the existing *Botanical Magazine* and *Botanical Register* and also 'beat them hollow by its cheapness and execution'. He would not urge anything more unless Neill approved, beyond suggesting that his magazine should be issued quarterly since this would reduce costs of carriage. He would gladly supervise the colouring of the figures – for Hooker was an accomplished botanical artist – and take care that they should be done as cheaply and as well as possible.[36] He had already experienced one disappointment, for the previous year he had suggested to Archibald Constable the idea of publishing a botanical periodical, but Constable had turned this down since there would be insufficient demand to make it profitable. This he did so very politely, since any work of Hooker's 'was bound to be creditable'. Neill appeared to be acting as broker between Hooker and Constable. In October 1822 Neill had passed onto the latter a letter containing Hooker's proposal to publish an English version of Linnaeus' *Species Plantarum*. This also was turned down. No doubt it would be a useful publication but Constable feared the study of botany had not yet become so general as to justify such a major undertaking. In any case, given the state of the country, he would not contemplate embarking on any publication with so many volumes, even on topics of general interest.[37]

From these early business contacts a warm friendship developed between Neill and Hooker, which lasted the rest of Neill's life.[38] He visited Hooker's home both in Glasgow and in London. He became a friend of the family and took an interest in the development of the Hooker children, especially Joseph Hooker whom he first got to know when he was only a few years old. Almost all the available correspondence refers to letters from Neill about plants: what he was growing, what he was sending Hooker, what he had recently received from him and so forth and these are summarised in the account of Neill's botanical interests. But sometimes there were little bits of gossip or comment which reveal the warm relations between the two men. Thus in 1830 Neill sent Hooker a specimen of an *Alstromeria* species, together with a drawing of it by Dr Greville and a *Portulaca* grown from seed he had received from Dr Gillies who was collecting in South America. He signed off with 'best wishes to Mrs Hooker and the youngsters'. The following year he again wrote about plants, but also mentioned that the Chair of Agriculture at Edinburgh University had become vacant. The salary was only £50 a year while the fee per student was restricted by deed of mortification to one guinea per student. Since the number of students would be unlikely to exceed 50 the total income was just £100. This was the vacancy mentioned earlier that John Fleming regretted not having heard about in time to enable him to apply.

In 1831 Neill was recounting a recent domestic alarm. The daughter of his cousin at Haddington, while staying at Canonmills was suddenly taken ill with 'the primary symptoms' of 'coolness of the extremities'. Medical advice and basins of warm water were administered and after 20 hours the symptoms passed. In ordinary times Neill would have thought they were due to her having eaten something which disagreed with her but 'in such times' the symptoms were alarming. This very likely refers to a cholera epidemic, which would cause anxiety over any indisposition. He mentioned this experience so that Mrs Hooker need not be alarmed if one of her family should develop similar symptoms.

In July 1832 he wrote to Hooker about the affairs of the Caledonian Horticultural Society, particularly how successful it had proved – especially in the establishment of the Experimental Garden. But this had led to an increase in costs, so that outstanding arrears due from members had to be called in. His records showed that Hooker had been admitted to the society on December 5 1822 and that he currently owed the society nine guineas.

In September 1832 Neill hoped that Hooker would visit Canonmills and also look at some plant novelties at James Cunningham's Comely Bank nursery. This was the Cunningham who first hybridised rhododendrons. In September 1834 Neill enquired about a German article that spoke highly of the statistical studies of his friend Dr Cleland of Glasgow. He thought the article comprised a translation by Hooker of some of Cleland's work. He could not find the journal in the local bookshops but would be grateful if Hooker could indicate how he might procure a copy, for 'I relish such things'. Later the same month he thanked Hooker for a copy of the translation from the Dorpel Annalen, with the favourable comments about Cleland. He had to report the sad news that their mutual friend Dr Gillies had died suddenly, apparently of a fit. He had taken a great interest in politics and Neill thought that recent political developments had hastened his end. He informed Hooker of a likely visit within the next two or three weeks from James McNab, who would be en route for his American plant hunting expedition. Neill bid Hooker 'waft him across the Atlantic at this season'. Audubon was in Edinburgh promoting the latest volume of his ornithological work. In a couple of year's time he was due for another expedition in America, likely to yield a rich return in both bird and plant specimens. He then added that it was desirable that either Mr Macgillivray (Robert Jameson's assistant) or James McNab should accompany him. This suggests that Neill may have been campaigning for such a joint expedition, although if so in vain.

In October 1836 he drew attention to Dr Gilbert McNab – another son of William McNab. He was glad to hear that Hooker had heard of his work and wrote: 'He will be well deserving of anything you can do for him, and will justify the highest character you can give him for steadfastness, thoroughgoing knowledge of his profession as a surgeon or as a physician, and for botanical zeal.' Gilbert McNab (1815–1859) went to Jamaica in 1838 and stayed there the rest of his life. Neill ended his letter with the hope that Sir William – for he was knighted by then – and Lady Hooker would visit Canonmills, where as well as the old inmates, he had six psittacinae (i.e. parrots), as well as some new plants.

In 1842 Neill sought Hooker's advice about the proposed bust and testimonial in honour of Professor Jameson. He enclosed a draft proposal to be sent to potential subscribers. Evidently he and Dr Traill had the task of organising the honour. They wanted the testimonial endorsed by his friends but were at a loss how best to proceed. Should they lay the proposal before professors at Oxford and Cambridge? Should it be brought to the notice of the Royal Society, the Linnaean and the Geological Societies of London? Dr Forbes had thought this might be appropriate but suggested that Neill should seek the opinion both Hooker and Robert Brown. Neill was concerned about doing anything that might hurt Jameson's feelings. Was there any precedent for this kind of action? Could it be carried out with delicacy in an unobtrusive way? Hooker's opinion would be 'very obliging' to Dr Traill and himself. This letter contained the current list of subscribers to Jameson's bust.

In October 1843 Neill was expressing his satisfaction at seeing Hooker's name on a list of subscribers to a testimonial in honour of William McNab. He added: 'You know the merits of the man; and I know the influence of your name as patron. It is therefore unnecessary for me to enlarge on the topic. Your advice on the best mode of proceeding will be most acceptable.' He later acknowledged Hooker's donation of two guineas.

On June 3 1844 after comments on James Drummond's seeds collected in Australia, he ended his letter with a wish to be remembered to Hooker's father, 'my excellent friend who enjoyed so much my mistaking his Goliath gooseberries for red magnum plums'. This little episode of mistaken identity became a sort of family joke. In the autumn of the same year he acknowledged the heavy task of correspondence with about 40 candidates for prizes at the fruit and dahlia show of the Caledonian Horticultural Society. He expressed concern at Professor Graham's failing health which had not been helped by his great concern over the loss of some of the Australian collections. The letters of the next few years were entirely botanical in nature and are therefore dealt with under that heading.

In 1845 on the death of Professor Graham, the Chair of Botany at Edinburgh University fell vacant, together with the curatorship of the Botanic Garden, which went with the honorary post of Queen's Botanist. Joseph Hooker decided to apply for the Chair. The appointment was complicated by the need to appoint the curator at the same time. While the appointment of the professor was the town council's responsibility, the curatorship was in the hands of the Lord Advocate, as the Government representative. It was obviously desirable that the same man should hold both posts but this was not guaranteed. If the Lord Advocate moved independently to appoint someone to the Botanic Garden this would have repercussions for the council. Indeed at one council meeting it was rumoured that the Lord Advocate had already appointed to the Garden one of the candidates for the Chair, thereby making it difficult for the council to appoint anyone else. However, the Lord Provost was able to scotch this rumour.

As soon as Neill knew that Joseph Hooker was interested in the Chair, he started campaigning in his favour. At times he was frustrated by Hooker's failure to appreciate how much canvassing he had to undertake to impress a sufficient number of the councillors, so he advised him how best to proceed. He should make a point of calling on each councillor, conveying the flattering impression of their key role in the appointment –just as Professor Graham had done before he was appointed. He must formally apply for the Chair in a letter to the Lord Provost. He must amass a formidable sheaf of testimonials. Hooker seemed to think that a few references from distinguished botanists would be sufficient. He was rather reluctant to sing his own praises and face the drudgery of gathering testimonials. Neill knew from past experience that this was wholly inadequate, for his principal rival for the job – John Hutton Balfour, the Professor of Botany at Glasgow University – was likely to submit 60 or 70 testimonials. Neill urged Hooker to get references from influential people like Sir James Ross, Sir H. Ellis, the Bishop of London etc. Even testimonials from former teachers were worth including, for it was quantity that impressed the councillors. Neill scurried between the Lord Provost and the principal of the University – who seemed sympathetic to Hooker, flourishing the first volume of Hooker's botanical account of his Antarctic voyages as surgeon naturalist with Sir James Ross. Neill wrote to Joseph Hooker's father William about the council and its ways. There was a decided majority of radicals, dissenters and enemies of any Church establishment, with Lord Provost Black at their head. But he had an independent spirit and prided himself, as a matter of principle, on how he dispensed the patronage associated with his office. He claimed always to prefer the most deserving, even though he might be a Tory or a churchman. A considerable number of the councillors generally voted with the Chair, although others liked to judge for themselves. Many would be ignorant of the scientific standing of the botanists among Hooker's supporters. He instanced the president of the Royal Society of Edinburgh, who on the occasion of a former vacancy for the Chair of Botany had asked: 'Who is this Robert Brown?'

During August of that year letters flowed between Neill and the Hookers, relaying the state of play, the latest rumours Neill had picked up, or advising on the next move. Balfour had been in Edinburgh enquiring about the appointment to the Garden. He had been to see the Lord Advocate and had not resigned the Glasgow Chair. He appeared to be waiting to see what happened about the Garden appointment. It was possible that if he were not assured of that he would stay in Glasgow. Speculation was rife.

Ultimately Hooker won testimonials from a glittering array of the leading botanists throughout Europe and these were printed and bound by Neill and Company for distribution. At one point Hooker let it be known that he was not interested in being appointed to the Garden alone or in the Glasgow Chair. Neill thought this very bad tactics.

Several men applied for the Chair but later withdrew, leaving Balfour and Dr William Seller (an Edinburgh M. D. and botanist) and Hooker as the rival candidates. The town council was split between the Whig and Tory factions that would often oppose each other in their support of rival candidates. Seller's chances for a while looked strong since he would get the Tory support, but at the last stage he withdrew his application.

At last on September 23 1845 the council met to decide between Balfour and Hooker. The proceedings were fully reported in *The Caledonian Mercury* of September 25th. It was the Lord Provost's prerogative to take the first step, by recommending a candidate and he chose Hooker, citing his impressive list of supporters. He introduced his decision with a long preamble in which he reflected on his invidious position in making such a choice, particularly since one of the candidates – i.e. Balfour – had been a friend since boyhood days. But he had to exclude any hint of favouritism. At that point Balfour's supporters sprang into action. Their main argument against Hooker was that the array of prestigious support was due to the fame of his father. Hooker's star might be rising but Balfour's had already risen. He could have done just as well as Hooker if he had gone on Ross's expedition. His father had got him the job. After all, 'it was easy to swim when the head was held up' (this comment resulted in laughter). Other critics complained that Hooker was a taxonomist and lacked experience in plant physiology and medical applications. Hooker's testimonials lacked any reference to his teaching abilities, whereas Balfour was known to be a capable and popular lecturer. When the issue was put to the vote Balfour easily carried the day with 23 votes in his favour against 10 for Hooker, so he was appointed to the Chair and later the curatorship of the Garden.

In retrospect, failure to get the Edinburgh Chair was a fortunate event in Joseph Hooker's career, since instead of being tied to the rather mundane task of teaching botany to medical students, he was free to undertake his celebrated Himalayan expedition which led to the introduction of so many new species of rhododendron and his lasting fame.

Three years later in June 1848 Neill acknowledged to William Hooker that he had received the first part of a copy of Joseph's recent travels. Considering that it was only the previous August that he had parted from him in the Botanical Garden, it was almost magical to read so soon of his travels to Malta, Egypt, the Red Sea, Ceylon and India, as far as Dr Wallich's famous garden at Calcutta. He had been with him in imagination and was delighted with his vivid descriptions and 'offhand' style of writing. Later the same year Neill had the sad duty of reporting the death of William McNab. They were anxious that his successor should be on good terms with the superintendent of the Caledonian

Horticultural Society's Experimental Garden. There followed a cryptic comment on the impossibility of that happening if the Lord Advocate – responsible for the appointment – was influenced by certain individuals who were hostile to the Society and had vilified the prize committee as cheats. What lay behind this aside remains unknown but suggests some contention within the horticultural community. At McNab's funeral Neill had conferred with Professor Balfour and learned that he had his eye on James McNab, the director of the Experimental Garden. Neill had no doubt he was the most suitable person to succeed his father. If appointed it would be a great loss to the society but he would not stand in his way. The present clerk – Mr Evans – was bred a gardener, had working experience, knew all the rules and procedures and during his four years service had got to know the members, was of agreeable manner and could fill the vacant post. It was likely that Hooker's advice on the appointment would be sought. If he felt at liberty to recommend the proposed arrangement he would be doing a favour to Professor Balfour, as well as the prize and garden committee of the society. As so often the case, Neill was not passing up any opportunity to make the most of any influence he could count on.

In April 1850 he had diffidently to note that council members of the Horticultural Society were mortified to hear local nurserymen boasting at the receipt of rhododendron seeds from Kew. They were well equipped in the Experimental Garden to raise such seeds, for Mr Evans was a very able cultivator. The members of the society included most of the landed proprietors with a taste for horticulture, well able to appreciate the fruits of Joseph Hooker's collecting. It was regular practice to distribute seeds of newly-introduced species to them. It would be a great favour to the society if a few seeds could be spared. A few days later he received a packet of seed and hastened to apologize for jumping to conclusions and over-stating their concerns about the rhododendron seeds. He was well aware of how much care and attention had to be paid to the distribution of such material. If there were any duplicates in the Experimental Garden that might be of interest to Kew they would be gladly sent.

On October 23 1850 Neill received a letter from William Hooker containing a note from his son Joseph accompanying seeds of *Neillia* – the genus of Himalayan shrubs which had been named after him by David Don. They gratified him to an extent he could not easily express. He never expected to live to see a mature plant but he might see some inch-high seedlings. He then wrote that the paralytic stroke he had experienced the preceding August had left him very weak but still with the power of moving the fingers and thumb of his right hand. He was able to visit his greenhouse and hot-house every tolerable day and 'never did my plants please me so much, my present gardener being most successful as a cultivator'. He did not know Joseph Hooker's address but he asked Sir William to express his sincere gratitude. After sending regards to Lady Hooker, he signed off 'Dear Sir William, With much esteem, Yours Pat Neill'. That was the end of their correspondence of some 30 years for Neill died the following year.

William Jardine (1800–1874) was born in Edinburgh in 1800. He was the son and heir of Sir Alexander Jardine, the Sixth Baronet of Applegarth, Dumfriesshire.[39] Educated partly at home, at Edinburgh High School and at York, he entered Edinburgh University at the age of 17 to study medicine. He was taught Natural History by Robert Jameson, who later became a benevolent friend who lent him specimens from the University Museum – a rare privilege. He was taught botany by a private tutor. Although he completed the university course he never graduated, for in 1820 he married Jane Home Lizars, the daughter of the Edinburgh engraver Daniel Lizars, who was a friend of Neill. The newly wed couple took off for Paris where Jardine continued his medical and natural history studies, provided with a letter of introduction from Neill to A. Royer of the *Jardin du Roi*, whom Neill no doubt got to know during his earlier Caledonian Horticultural Society tour of gardens in northern Europe. However, the French sojourn was cut short when his father died in April 1821 and Jardine had to return home to Jardine Hall to take up his numerous responsibilities as Seventh Baronet.

From an early age Jardine was particularly interested in ornithology. While at university he became friendly with James Wilson, the janitor who was responsible for preparing and preserving the museum specimens. When Wilson died Jardine purchased his bird collection which became the foundation for an immense and internationally recognised collection of birds from all over the world. He was also a very competent artist and was able to draw, paint in watercolour and oil, etch and engrave in both copper and wood blocks. He was taught to draw by Patrick Syme, artist to the Wernerian Society. His great friend and collaborator was Prideaux John Selby of Twizell Hall in Northumberland, also an ornithologist, artist and illustrator.

While he was in Paris, Jardine had the distinction – remarkable for one so young – of being elected to the Wernerian Natural History Society and in 1821 read a paper on a new species of pigeon from New Holland. Jardine soon established a reputation as a naturalist. In 1834 he went on a trip to Sutherland to investigate the natural history of that little-known area. He was accompanied by his brother John, Prideaux John Selby, the botanist Dr Robert Greville and the entomologist James Wilson – another of Neill's Edinburgh friends. Jardine's first work on natural history – jointly with Selby – was the *Illustrations of Ornithology* (1826–1843). Jardine went on to do more than any other single person to popularize natural history among the reading public. He was the driving force behind the 40 volumes of *The Naturalist's Library* and the popular *Magazine of Zoology and Botany* – the early volumes of which were printed by Neill and Company. In 1826 both Jardine and Selby partook in a couple of drawing lessons from the American ornithologist and bird painter Audubon when he was visiting Edinburgh. Audubon was introduced to William Lizars by Neill and was so impressed that he immediately summoned Jardine and Selby to come and meet him.[40] Up to that point Audubon had enjoyed little success in winning recognition for his work. It is no exaggeration to recognise Neill's initiative in arranging this introduction as the turning point in Audubon's career in Edinburgh, for his fortunes immediately improved and he was soon being lionised.

Jardine had both a prodigious capacity for work and wide ranging interests, so characteristic of the Victorian era. In addition to his immense scientific correspondence, he ran his estate with a dozen farms, was a keen devotee of fishing and shooting, entertained his neighbours and was a member of the General Assembly of the Church of Scotland.

Neill's friendship with Jardine dated from the beginning of the latter's career when he enjoyed benevolent encouragement from the older man. The friendship continued throughout Neill's life. They would have been in regular contact via the Wernerian Society – of which Jardine was an active member and later vice-president – as well as the occasions when Neill and Company were printing for Jardine. Neill visited Jardine Hall as a guest. The cordial relations between them are illustrated in the occasional, surviving letters on different topics.[41]

In January 1822 Neill – commenting on proofs from a printing job he was doing for Jardine and Selby – was questioning whether a particular author's name should have one or two 't' letters. Jardine's spelling was notoriously idiosyncratic so Neill was probably right to question it. He also asked that appropriate indications should be inserted to help the bookbinders.

On September 9 1822 Neill thanked Jardine for his invitation to accompany Mr Syme and himself on a short expedition, although to where and for what purpose was not mentioned. Neill had to decline since George IV's visit to Edinburgh had quite upset his plans. Also, he had to complete a piece of executory business which could take him to either London or Belfast. He had sent a memorandum via Jardine to Patrick Syme, who was making arrangements for the journey. He had once had the pleasure of accompanying him on an excursion and greatly regretted being unable to do so on this occasion. He referred to an apple which Jardine had sent him for comment. Neill thought both its appearance and scent were promising. It was evidently close to the ribston pippin, of which it was probably a seedling. The fruit would continue to improve for two or three years. He had exhibited the apple at the general meeting of the Caledonian Horticultural Society, where the hope was expressed that Jardine would send in two or three specimens the following autumn. Neill thought it would be a good idea to take two or three shoots from the tree and graft them on to wall trees. He concluded his letter with the observation that Mr Adie – the Edinburgh optician – had a new apple in his garden, derived from seed he had received from Jardine's father. It was a baking variety, unlike Jardine's apple, which promised to become a favourite dessert fruit.

In July 1830 Neill wrote to Jardine asking for his assistance in straightening out a question of priority. In August 1816 Neill had visited the Rinns of Galloway to look at the fishponds. The following month he had published a full account in *The Scots Magazine* of the ponds and their inhabitants. In addition, in the article on icthyology in Brewster's *Edinburgh Encyclopaedia*, Dr Fleming had included Neill's account with acknowledgement. Neill recollected some two or three years previously, that *The Scotsman* newspaper had printed an excerpt from the Dumfries paper and had commented on the novelty of the ponds. However he had just seen *Sketches from Nature* by Mr Jack Macdiarmid who had visited the ponds in 1824 – eight years after Neill's account *in The Scots Magazine* and seven years after the report in the *Encyclopaedia* and yet made no mention of his article. Perhaps Macdiarmid was not aware of it and if so, could Sir William put him straight on that score. If he was indeed unaware of Neill's precedence 'let him pass'.

Later the same year, Neill referred again to the ponds and recalled that he had gone on from there to the Isle of Man where, while in the Botanic Garden early in May, he had heard a 'merry peal' rung out to mark the marriage of Princess Charlotte of Wales. He seems rather to have relented over his irritation with Macdiarmid for he added that his papers were very clear and accurate and that he had no need to fear doing justice to any who had priority.

In September 1833 Neill wrote to Jardine thanking him for the specimens of *Saxifraga hirculus* from its Scottish habitat. This was a new discovery since it was not mentioned in Hooker's *Flora Scotica*. Also neither David Don from London, nor James T. Mackay from Dublin – two of his botanical friends mentioned earlier who were then staying with him – were familiar with the species. Neill then added: 'It will give me great pleasure to find that I can add to your collection of plants.' It is likely this comment referred to the practice among collectors of exchanging dried specimens of the native flora. He ended his letter with a few comments on proof corrections.

In November 1835 Neill acknowledged a letter from Jardine and congratulated him on his return to his paternal seat, which he had had to vacate for a period. How quickly the three and a half years had passed. He recalled being at Lochmaben when Jardine was preparing to leave and yet it only seemed a few months ago. He was delighted to learn that Jardine's *Pinus Douglasi* (*P. menziesii*) was thriving on a mossy site and had reached ten feet in height. Jardine's specimen must be next in size to Sir John Leslie's, which Neill had found to be eight feet tall in August 1832, shortly before Sir John's death. His successor had recently told Neill that it was now 12 feet, growing on loamy soil. Jardine had expressed a desire to obtain some American plants, for he was now adding gardening to his list of interests. If he knew what was wanted, Neill would enquire of Mr Barnet – the gardener in charge of the Horticultural Society's Experimental Garden – to see what might be available. His own garden was too 'circumscribed' to provide what Jardine wanted. He expressed his gratitude to Jardine for gifts of plants. He had received *Nierembergia phoenicea* before it had been figured in London. Species of that genus which he had raised at Canonmills were in great demand in the south, especially *N. intermedia, gracilis* and *calycina,* which had first flowered in this country at Glasgow. The rest of this letter was taken up with his interest in persuading Sir Charles Bell to return to Edinburgh to take up the recently vacant Chair of Surgery.

In January 1837 Neill was thanking Jardine for his presentation to the Wernerian Society library of a copy of his 1836 address to the Berwickshire Naturalists' Club. It is rather amusing to note that, in this formal note sent as secretary of the Society, he signed himself, 'Your very obedient Servant' in place of his usual, 'Dear Sir William, Yours truly' or 'With esteem' etc.

Later the same year he sent Jardine all the information he had about snowy owls, culled from his *Adversaria* or from his own recollection and that of Mr Lawson. The delay in his reply to Jardine's enquiry was due to pressure of public business, i.e. his duties as a member of the town council. He added: 'I need scarcely add that I would much rather write about snowy owls etc than wrangle with town councillors.'

A little later he was asking Jardine to procure a specimen of the vendace fish for the benefit of a Mrs Lee who wished to make a painting of it. The vendace was a species of whitefish –

widespread in Northern Europe – with a local form in the lochs about Lochmaben. At one time many angling clubs were devoted to its capture.

In December 1838 he responded to Jardine's identification of a gull that Neill had sent him for comment. Jardine thought it was an ivory gull but Neill was unconvinced. He had consulted all the books he had to compare the diagnostic characters of the species and still stuck to his original identification of the specimen as an Iceland gull. He asked Jardine to confirm whether he may have misinterpreted his meaning and also which characters he had used for identification. There was no doubt that the legs were not brownish black in colour, but rather flesh coloured so he asked him to look again at the specimen. Although Jardine was a recognised authority, Neill was not prepared simply to accept his identification without question. These few letters are sufficient to show the kind of easy relationship that existed between the two men based on their shared interests in all branches of natural history.

Walter Calverly Trevelyan (1797–1879) inherited the Northumberland estate of Wallington and the baronetcy when his father died. Trevelyan was a man of great wealth who travelled leisurely about Britain and Europe with his wife, in pursuit of his scientific interests. Although primarily known as a geologist he appeared just as interested in botany, collecting plants for his herbarium and seeds and specimens for his botanical friends – who included Neill. The surviving letters from Neill to Trevelyan are largely devoted to comments about wild species from Britain or the European mainland, but often also refer to what Neill had been doing or whom he had recently met. Such personal information has been extracted for notice here.[42] The strictly botanical information has already been noted.

The first example of the sustained correspondence between Neill and Trevelyan dates from September 1826. Trevelyan was about to embark on a visit to northern Europe and had consulted Neill about persons he should arrange to meet, since he would have known that Neill was well acquainted with the horticultural scene in Flanders and northern France from his visits there in 1817 and 1821. Neill obliged with letters of introduction to M. Royer (who spoke English fluently) and the Mayor of Enghien. He reminded Trevelyan that no letters must be sealed. M. Royer was reserved but most friendly and helpful when his confidence was won. If he saw a M. Bose, could he tell him they were just beginning the Experimental Garden at Edinburgh. Another person to contact – M. Parmentier – was a frank, plain burgomaster who would readily interrupt his duties to greet a botanist or talk about plants. He advised Trevelyan to visit the Duc d'Arenberg's park which was quite near Enghien. Neill also sent along a letter to Mr Blaikie, who was now back in royal favour. It was quite a treat to hear Blaikie recounting his adventures in Jacobin times or having him point out the scenes of remarkable events. He also had a letter for Dr Sparzheim from Mr Combe.

In March 1832 Neill sent Tevelyan an extract from the minutes of the Caledonian Horticultural Society, thanking him for his donation to the library of a small, scarce volume entitled *The Mystery of Vintners* (London 1675) by Dr Charleston. He hoped it would afford a good example for others to follow.

In January 1834 Neill commented on the unusually early spring and then referred to a discussion he and Trevelyan had been having about the persistence of viable seeds in old soil.

Trevelyan had carried out an experiment but Neill thought he had not been 'nice enough' about the way he did it. He needed to take a deeper sample of soil, exclude it from contact with the air, cover it with a close fitting bell jar and water with distilled water. Neill had taken a sample of earth from the lowest layer (containing oyster shells from the foundations of Sir W. Forbes' new bank), using great precaution except for watering. No herbaceous plants germinated, although there was plenty of conferva or perhaps germinating moss, similar to what appears on pots in the greenhouse and attributed to contamination from the air. He thanked Trevelyan's father for the present of nonpareil and margil apples from their Somerset orchard.

A few days later Neill wrote again to seek a favour, prompted by Trevelyan's 'obliging disposition'. Neill had lost his mistletoe the previous summer due to canker on the host apple tree and had been unable to find a replacement. He wondered whether Treveyan could find a Somerset nurseryman who could supply mistletoe growing on a transportable thorn, apple or pear stock which could be dug up, packed in damp grass and moss and sent by wagon to London and thence by steamer or smack to Leith, suitably addressed. Of course he would pay for it. Thanking again for the present of apples, he noted that the fruit from the standard trees had a higher flavour than those from the wall. He ended his letter with a note about the Wernerian Society, which was doing well. The zoologists had been puzzled by the discovery in the Union Canal of a species of mussel (*Mytilus*) native to the Danube and apparently imported with undressed timber. Neill was keeping some of them in his aquarium in the greenhouse.

In April 1834, along with botanical comments, Neill reported that James McNab was shortly to leave for a plant hunting expedition to North America, hoping to cover his expenses by providing dried or live specimens. If anyone wanted to subscribe there was no time to lose in doing so. McNab was an excellent, practical botanist and a very honourable young man. Neill reported that his gardener, William Brackenridge, should have arrived by then in the Berlin Garden. He had received so good an offer that he could not do otherwise than advise him to accept. Brackenridge – a Scot – later emigrated to America, served as assistant naturalist in Wilkes' United States Exploring Expedition (1838–1842), became superintendent of the National Botanic Garden at Washington and finally, nurseryman at Govanstown, Maryland.

In June 1834 Neill was in London for a few days and wrote to Trevelyan from the Craven Hotel on The Strand. He had learnt from David Don that Trevelyan was also in London, and so he hoped to see him. Neill had visited the Zoological Society the previous day. He was troubled by the heat which 'does annoy me'. He found it much 'closer' than Edinburgh, compelling him to be cautious and not overwork himself. Perhaps he was a mild hypochondriac. He reported that he had received a visit at Canonmills recently from Trevelyan's brother Arthur, accompanied by Mrs Grant of Monymusk and another lady.

In March 1835 Neill provided Trevelyan with information about the Caledonian Horticultural Society. Members could introduce their friends to the Experimental Garden. The cost of membership was in the form of a share costing £21, although he had heard shares could be bought for £14 or £15 on the market. Because the garden was low in funds, members had to pay for any plants they wanted, although at a lower rate than for non-members. However, members were entitled to free grafts and seeds. When they were out of debt the former provision of free plants would be resumed.

In September of the same year Neill had just returned from a visit to Ireland, having apparently attended a botanical meeting in Dublin (since he mentioned in a letter to Trevelyan several botanists who were also present or expected to attend). Trevelyan was then in Jersey and was about to embark on a continental tour. He had recently wed so this was probably his honeymoon. Neill wished him much happiness, hoped he would steer clear of the cholera and find Old England safe on his return, for 'I do feel we are in a sad plight'. This could well have been a reference to the passage of the Municipal Corporation Act which revolutionised local government in England. It is likely that Neill was no devotee of reform. He had also met Combe – the Edinburgh phrenologist – in Dublin and had inspected the preserved cranium of Dean Jonathan Swift. He detected unnatural lumps on the inside and other signs of abnormality. Recalling that Swift survived during the last two or three years of his life in a state of idiocy, Neill took such signs as compelling evidence in support of phrenology and likely to confound the scoffers. David Don – the botanist son of George Don – had also been in Dublin. They stayed in the same house and travelled back to Edinburgh together. Don was now staying with Neill at Canonmills. Don was a candidate for the Chair of Botany at King's College. Backed by Robert Brown, it is perhaps not surprising that he got the job. Neill added a note that he had been to see a large working model of Mr Stein's steam engine. Generating one horse power and about the size of a tea kettle, it did all that might be expected of it.

In June 1840 Trevelyan was about to leave for a visit to Arran. After acknowledging receipt of plant specimens which Trevelyan had collected in The Faroes and suggesting a few he would very much like to have from Arran, Neill passed on some advice about places to visit. Trevelyan should visit the Witch's Height or Top, Ceim-na-Caillich and the little harbour of Drumlaborach, where a trap dyke formed a natural jetty or pier. Even more importantly he should visit South End Harbour, which was a naturally formed harbour with its breakwater and piers as if built by man, yet all natural dykes, traversed by red sandstone and greenish limestone. The many dykes on one side of Brodick Bay had surprised him, as well as the plants he had found in Glen Rosa. He noted there was a paper on Arran geology in the press by Necker, who had described several hundred dykes, with sketches of their location.

On January 10 1843 Neill wrote to Trevelyan to thank him for the present of seeds recently collected in Greece. He had asked McNab to divide them into three packets: one for Professor Graham and the Botanic Garden, one for the Society's Experimental Garden and the third for himself. They were all very grateful for his 'obliging attention'.

The Newhaven Railway Company had sadly annoyed him. Their deep tunnel had cut off all his spring water, had dried up the pond where his tame cormorants were wont to dive and to crown all, they took advantage of his absence in London to fill up Canonmills Loch with the aid of 150 Irish labourers who worked all night, so that no interdict could be lodged in time. He concluded his letter by noting that the bust of Professor Jameson was well advanced and also that the Caledonian Horticultural Society had insisted that he should also sit for a bust by Mr Steele, the Edinburgh sculptor.

In the spring of the same year he acknowledged receipt of a small box containing specimens of *Testacella halotidea* (carnivorous shelled slugs which prey on earthworms). If Trevelyan had put some moss in the box, four might have survived instead of just two. Having ascertained their habits by consulting his *Dictionnaire d' Histoire Naturelle*, he put them in a pot supplied with earth and two worms. The following morning he found that one had descended into the earth while the other was attacking one of the earthworms, which was wriggling in an effort to escape. He reflected what a boon these animals would be to collections of seedling plants if they would attack slugs but he feared they would ignore their relatives in favour of the 'beefy worms'.

In March of the following year he mentioned that the Caledonian Society's exhibition hall would have on display some new and rare flowering plants of his own – but trifles, he added modestly. He then wrote about a short tour he had once made in Suffolk and Norfolk, serving to remind us that he was also an amateur geologist. He was struck by the massive erosion due to the action of the sea and heavy rain, especially in the vicinity of Cromer, where his attention was drawn to an additional cause which seemed to have been overlooked. He observed that in dry weather a strong wind from the north east carried fine sand from the cliffs inland, loosening coarser fragments and fossils which fell to the base of the cliffs to accumulate to a depth of several feet. Where the cliff was more resistant, dense, or comprised of layers of clay, wind action was ineffective and hence such layers stood out in bold relief. He thought a high wind on a dry day could be more effective in eroding the cliff that when combined with rain, which would prevent the sand being blown away and would not act (like the wind) on the entire cliff face. He remarked on a former promontory which had once extended two miles from the coast and had now vanished. The removal of such a feature might be expected to alter the direction of currents acting on the coastline.

In January 1845 there is mention of Neill's decline in health. After commenting on Professor Forbes' zeal as secretary of the Royal Society of Edinburgh and Professor Graham's serious illness, he mentioned that he had been unable to attend two recent Botanical Society meetings. They were held in the evening and he was forced to be absent because he was now 'unfit for going out at night'. His real reason for writing was to solicit a short communication from Trevelyan to be read at the next meeting of the Wernerian Society.

The following year Trevelyan was in Portugal and had written to Neill from Lisbon, with comments on his plant collecting and a present of seeds. Neill replied that he had read extracts from Trevelyan's letter to the Botanical Society. Also, he had heard from Trevelyan's brother that their father was too unwell to write. Neill wondered whether the Portugal trip should be abridged. He noted that this year he was a member of the General Assembly of the Church of Scotland and had been attending the meetings to the point of exhaustion. He found that late hours did not agree with one who had entered his 70th year. Professor Balfour had a large class of about 150 students and was very popular. He hoped he would set about describing the species that first flowered at Edinburgh, like his predecessor – a hope which was not fulfilled.

In July of the following year he informed Trevelyan of his interest in the edible snail, *Helix pomatia*. He had kept them in his garden for many years until his heron extirpated them. He now had a separate enclosure for the herons and gulls so he could consider keeping snails again. He wondered whether they were naturalized with Trevelyan, no doubt hoping to profit thereby if they were. As feared, Trevelyan's father's illness had proved fatal so Neill's letter was on black edged notepaper.

Early in 1848 Neill thanked Trevelyan for a sample of lemons grown at Wallington. He would exhibit them at the next meeting of the Horticultural Society. It would be of interest to know how the tree was grown – whether in a conservatory or otherwise – as well as its age and general management. He remarked that Professor Jameson was remarkably well for one in his 77th year, while he himself was not doing too badly in his 73rd, apart from his failing eyesight. Now that Newcastle was within four hours of Edinburgh and noting that Trevelyan often wrote from Wallington – not far from Newcastle – he hoped to see him and his wife at 'his little spot' in Edinburgh. Trevelyan had inherited the baronetcy so Neill now signed off: 'Dear Sir Walter, very truly yours.'

In November 1849 Trevelyan wrote on behalf of friends for advice on the construction of a conservatory and heated greenhouse. Since the project was quite elaborate, Neill suggested that the best solution was to approach James McNab, who in his opinion was the best person in Britain to provide technical advice. Ever cautious, Neill added a P.S. to the effect that Sir Charles Smith, Queen Street, Edinburgh, was the principal professional planner and garden architect in Scotland. He did not in any way mean to disparage Smith by naming McNab as the man whose opinion he would follow. Neill's advice was taken and McNab provided a detailed plan.

On February 9 1850 Neill had an interesting comment about John Jeffrey – the Edinburgh Botanic gardener from Perthshire – who was being considered for the Oregon Association plant collecting expedition on the Pacific coast of North America. Neill described him as a very excellent botanist and a 'thorough-going' man who would not be daunted by difficulties and would collect not only seeds and plants, but also insects and minerals. The Oregon Association – based in Edinburgh – raised subscriptions from many of the landed gentry who hoped to obtain new species from Jeffrey's collection. Neill assured Trelawney that he would notify Lord Murray – who was organising subscriptions – if he wanted one or two shares in the expedition. Unfortunately subscribers had no conception of the difficulties Jeffrey faced and were unduly impatient for results. Although Jefrrey collected many new species, his reports were considered too infrequent and what he sent back less than expected so his contract was prematurely terminated. He was last heard of in 1854 either in San Francisco or Mexico, apparently having died under unknown circumstances – truly a tragic story, which does little credit to the Oregon Association.

In November of the same year Neill thanked Trevelyan for a box of game 'of the most varied description' which had just arrived. Mr and Mrs McNab were to dine with him the following day and thus share in his bounty. In January 1851 Neill informed Trelawney that the anniversary meeting of the Botanical Society was to be held at Canonmills – very likely out of deference to Neill's frail condition – and that he hoped for the pleasure of his company on February 8th at five o'clock in the afternoon. Not long afterwards he acknowledged a present of guavas grown at Wallington and a welcome addition to the Society's dessert. There the correspondence ended for Neill died in September of the same year.

1   Moray McLaren (ed), *The House of Neill 1749–1949*, 1949, Neill & Co, Edinburgh.
2   Patrick Neill, *Short Biographical Memoir of the Late Mr. John Mackay, Superintendent of the Royal Botanic Garden at Edinburgh*, *The Scots Magazine, 1804*, (66), pp.95–99.
3   Notes from Royal Botanic Garden Edinburgh, 1904, pp.21–49.
4   *Ibid*, pp.94–95.
5   G. C. Druce, *The Life and Works of George Don*, Notes from Royal Botanic Garden. Edinburgh, 1904, (3), pp.53–290.
6   J. Grant Roger, *George Don, 1764–1814*, *The Scottish Naturalist*, 1986, pp.97–108.
7   G. C. Druce, *The Life and Work of George Don*.
8   Patrick Neill, *Biographical Note of the Late Mr George Don of Forfar*, Trans. Bot. Sopc. Edinburgh, IV, 1850–1853, p.117.
9   MS. 2257, ff.103, 109–114, National Library of Scotland.
10 Neil Letters, The Linnean Society, London.
11 *Ibid*.
12 William Hooker's Correspondence with Patrick Neill, Library, Kew Gardens, London.
13 MS 2257, 103, 109–114, Letters Relating to Financial Assistance to Mrs Don, National Library of Scotland.
14 *Ibid*.
15 Proc. Linn. Soc., 1842, (130), pp.145–149.
16 *Gardeners Magazine*, 1829, p.534.
17 J. Bot., 1846, p.534.
18 Trans. Bot. Soc. Edinburgh, 1886, (16), pp.187–189.
19 Notes from the Royal Botanic Garden. Edinburgh, 1908, pp.293–324.
20 Trans. Bot. Soc. Edinburgh, 1979, pp.381–383.
21 Harold R. Fletcher and William H. Brown, *The Royal Botanic Garden Edinburgh 1670–1970*, 1970, pp.139–141, HMSO.
22 *James McNab's Scrapbook*, Library, Royal Botanic Garden, Edinburgh.
23 Trans. Bot. Soc. Edinburgh, 1866, (8), pp.463–476.
24 Trans. Bot. Soc. Edinburgh, 1857, (6), pp.16, 20–22.
25 Notes Rec. R. Soc., 1951, (9), pp.115–136.
26 Harold R. Fletcher and W. R. Brown, *The Royal Botanic Garden Edinburgh 1670–1970*, p.70.
27 John Keay and Julia Keay (eds), *Collins Encyclopedia of Scotland*, 2nd Edition, 2000, p.581.
28 Charles Gibbon, *The Life of George Combe, Author of 'The Constitution of Man'*, 1878, Volumes 1 & 2.
29 MS 7240, f.160, The National Library of Scotland.
30 John Keay and Jukia Keay (eds), *Collins Encyclopedia of Scotland*, p.103.
31 Trans. Bot. Soc. Edinburgh, 1860, (6), pp.119–128.
32 William Hooker's Correspondence with Patrick Neill, Library, Kew Gardens, London.
33 Trans. Bot. Soc. Edinburgh, 1868, (9), pp.414–426.
34 Proc. Linn. Soc., 1865–66, lxvi–lxxiii.
35 Eric W. Curtis, *The Story of Glasgow Botanic Garden*, 2006, p.30, Argyll Publishing.
36 William Hooker's Correspondence with Patrick Neill, Library, Kew Gardens, London.
37 MS. 791, f.638. National Library of Scotland.
38 William Hooker's Correspondence with Patrick Neill, Library, Kew Gardens, London.
39 Christine E. Jackson and Peter Davis, *Sir William Jardine. A Life in Natural History*, 2001, Leicester University Press.

40 John Chalmers, *Audubon in Edinburgh*, 2003, NMS Publishing, Edinburgh.
41 William Jardine Correspondence with Patrick Neill, Library, National Museums of Scotland.
42 Walter Calverly Trevelyan, Correspondence with Patrick Neill, Library, University of Newcastle-upon-Tyne.

10 ❧ End of Story

Neill's Will

By the standards of his times Neill enjoyed a comparatively long life. In his latter years he took care about his health and refrained from going out at night. He was latterly handicapped by failing eyesight but his interest in his plants never wavered, even when severely incapacitated by a stroke shortly before he died. His extremely long and detailed will provides confirmation of the friendships and relationships which he valued most during his life. Ann Neill – his cousin James' daughter – was the chief beneficiary and quite rightly so since she had run his household for so long. Apart from Neill's natural history books, she inherited Canonmills Cottage and contents, the garden, plants and the animals. He was really concerned that his plants and his little menagerie of animals should continue to be looked after. At her death, without issue, all the property concerned was to pass to his 'nameson' Patrick Neill Fraser, with the further provision that anyone who inherited the property thereafter should occupy it and take the name of Neill. This desire to perpetuate his name might suggest that lack of descendants may have cast a shadow on his life, acknowledged only in his will.

The other principal beneficiaries were his several printer and bookseller relatives in Haddington and their offspring. But there was also a long list of individuals who inherited £50, £100 or £200 – often with a qualification worth noting. Thus, he bequeathed £100 to 'my dear friend, James Townsend Mackay, Director of the Botanic Garden, Dublin', to purchase a natural history book. Likewise, to 'my esteemed Professor Jameson £100 to purchase a natural history book', to Professor Thomas Stewart Traill £100, and to 'my esteemed friend' Robert Bald, mining engineer, Alloa, £50. To Sir William Jackson Hooker, £50 to purchase a natural history book. To Dr John Fleming £200 or in event of predecease, to his widow and/or son. Neill's natural history books were to be divided between Fleming and Patrick Neill Fraser, with Fleming having first choice. Further, all his books in manuscript, other manuscripts, letters and loose papers – all the items whose loss we so much regret – were bequeathed to Dr Fleming. Several ministers were singled out for bequests of £50 each, including the Rev. James Grant, Minister of St Mary's Kirk, Edinburgh of which Neill was an elder. Among the several ladies who received bequests was Ann Patricia Neill McNab –

daughter of James McNab who, like Neill's partner William Fraser, had demonstrated his regard for Neill by naming one of his offspring after him. Robert Stevenson's daughter was also left £300. One beneficiary who arouses unsatisfied curiosity was Marianne Kerr of High Street, Burntisland, who was bequeathed 10 guineas to purchase a ring 'as a remembrancer'. Was this an echo of a romantic attachment?

Apart from the bequests to individuals there were also sums of money left to the various institutions which Neill supported. Chief among these were the bequests of £500 to both the Royal Society of Edinburgh and the Royal Caledonian Horticultural Society. The interest every second or third year was to be used to provide a medal or other award to respectively a distinguished Scottish naturalist and a distinguished Scottish botanist or cultivator. These two valuable bequests have guaranteed the survival of these prestigious awards to the present day. Other institutions that were awarded lesser sums included the School of Arts in Edinburgh, the Parish School of St Mary's, The British and Foreign Bible Society, The Royal Infirmary in Edinburgh and the General Assembly of the Church of Scotland, for religious and educational use. His association with Orkney was not overlooked, for his bequests included the Orkney and Zetland Society for relieving poor natives and sending them home. Neill drew income from feu duties inherited from his relatives in Stronsay. He directed his trustees to make over the greater part of them for the benefit of the local schoolmaster and poor children, at the discretion of the schoolmaster.

Thus, Neill's will displayed his earnest desire to demonstrate in a practical way his regard for his relatives, his closest friends and the institutions he supported. He must have devoted great care and attention to the long statement which summarises for posterity the relationships which determined the pattern of his life.

Neill was buried in Warriston Cemetery, not far from Canonmills Cottage. His headstone commemorated his zeal for literature, science, patriotism, benevolence and piety – an accurate evocation of the values by which he lived.

Articles by the author

(1997) *The Working Life of Scottish Gardeners between the Wars. Review of Scottish Culture, No. 10: 67–85.*

(2000) *The Gardens of Cullen House, 1760. Garden History, 26:2, 136–152.*

(2001) *James Sutherland's Hortus Medicus Edinburgensis, 1683. Garden History, 29:2, 121–151.*

(2003) *John Adam's Eighteenth Century Walled Garden at Blair. Garden History, 31:1, 48–66.*

(2004–5) *William Gibbs: Lord Seaforth's Gardener, 1782-1812. Review of Scottish Culture, No. 17: 10–21.*

(2006) *The Diary of a Midlothian Nineteenth Century Garden. Review of Scottish Culture, No. 18: 73–86.*

(2007) *A History of Apples in Scottish Orchards. Garden History, 35:1, 32–50.*

(2008) *Orchards, Fruit and Gardeners in Early Nineteenth Century Scotland. Review of Scottish Culture, No. 20: 41–56.*

(2009) *The Eighteenth Century Millers and their Garden at the Abbey of Holyrood. Journal of Scottish Local History, August 2009, 15–22.*

(2010) *Pears in Old Scottish Orchards. Review of Scottish Culture, No. 22: 1–17. The Janet Allen Prize Essay.*